Anatomy *and* PHYSIOLOGY

BEAUTY THERAPY BASICS

Helen McGuinness

D0177361

Hodder & Stoughton

A MEMBER OF THE HODDER HEADLINE GROUP

A catalogue record for this title is available from the British Library

ISBN 0 340 639431

First published 1995
Impression number   10 9 8 7 6 5 4 3 2
Year   1999 1998 1997 1996

Typeset by Wearset, Boldon, Tyne and Wear.
Printed in Great Britain for Hodder & Stoughton Educational, a division of Hodder Headline Plc, 338 Euston Road, London NW1 3BH by The Bath Press, Bath.

CONTENTS

ACKNOWLEDGEMENTS

I would like to extend my special thanks to my father for his great skill and patience in designing the illustrations, and to my husband Mark for his considerable help, support and understanding in the writing of this workbook.

I am greatly indebted to my friend and colleague Deirdre for her professional help and contributions throughout the preparation of the text. My gratitude also extends to all the students and staff in the Beauty Therapy Department at Southampton Technical College who have supported and encouraged me in the development of this book and without whom there may not have been an original need for a workbook.

PREFACE

NOTE TO STUDENTS – HOW TO USE THIS BOOK

This workbook is intended for candidates undertaking an NVQ training programme in Beauty Therapy at Levels 2 and 3 and provides the essential information on anatomy and physiology required at these levels. The contents of this book have been written in order to present the basics of anatomy and physiology in such a way that they are not only easy to understand, but are relevant to practical work carried out in the salon. The material has been designed to be interactive with tasks and self-assessment questions throughout, in order to assess overall understanding of the individual subject areas.

The aim of this workbook is not only to assist candidates in learning the principles of anatomy and physiology but also to generate portfolio evidence of underpinning knowledge for an NVQ in Beauty Therapy. It is divided into separate chapters and at the front of the book there is a study grid for Levels 2 and 3, which indicates the knowledge relevant to the unit or units to be studied. At the back of the workbook is a candidate competence record, designed to enable Assessors to authenticate the evidence produced by candidates whilst working through the book, and may be used towards verification.

ACCOMPANYING TEXTBOOK

You will find it helpful to use an accompanying textbook to extend your knowledge and give you any further information you may need. The recommended text to use is *The Science of Beauty Therapy* (2nd edition) by Ruth Bennett, published by Hodder & Stoughton.

Study grid
Anatomy and Physiology for NVQ Level 2

Underpinning knowledge	Chap	A01	A02	A03	A04	B04	B05	B06	B07	B08	B09	B10	B11	B12	B14	B13	B15	B16	B17	B18	B19
					Units												Additional				
The skin	1										•										
The hair	2													•	•						
The nail	3													•	•	•					
The skeletal system – major bones of skull and face	4											•									
The skeletal system – major bones of skull, neck and face	4										•										
The circulatory system – head and neck	7										•										
The lymphatic system – head and neck	8										•										

Study grid Anatomy and Physiology
Beauty Therapy Basics – NVQ Level 3

Underpinning knowledge	Chap	Core units												Additional			
		C01	C02	C03	B04	B17	C04	C05	C06	C07	D01	E01	E02	C08	C09	C10	D02
The skin	1	•					•				•	•	•	•	•		•
The hair	2										•						•
The skeletal system	4						•	•		•			•				
Joints	5							•		•							
Muscles – facial	6							•					•				
Muscles – body	6						•	•		•							
Circulatory system	7						•	•		•		•	•	•	•	•	
Lymphatic system	8						•	•		•		•	•				
Respiratory system	9									•		•					
Olfactory system	10											•	•				
The nervous system	11							•				•	•				
The endocrine system	12										•	•	•				•
The female breast	13						•	•				•	•				•
The digestive system	14						•					•	•				•
The renal system	15						•					•	•				

Chapter 1

THE SKIN

The skin is a very important organ to a therapist as it represents the common foundation for all practical treatments carried out in the salon; therefore an understanding of its structure and functions is essential for carrying out treatments safely and effectively.

A competent therapist needs to be able to:

* relate knowledge of the skin to the physical effects of treatments

By the end of this chapter you will be able to relate the following knowledge to your practical work carried out in the salon:

* The structure and function of the epidermis
* The structure and function of the dermis
* The structure and function of the subcutaneous layer
* The appendages of the skin and their functional significance
* The blood, lymphatic and nerve supply to the skin
* The functions of the skin
* Treatable and non-treatable skin diseases and disorders

Before looking at the structure of the skin, let's consider a few facts...

* The skin is a very large organ covering the whole of the body
* It varies in thickness on different parts of the body. It is thinnest on the lips and eyelids, which must be light and flexible, and thickest on the soles of the feet and palms of the hands where friction is needed for gripping
* As the skin is the external covering of the body, it can be easily irritated and damaged and certain symptoms of disease and disorders may occur
* Each client's skin varies in colour, texture, and sensitivity and it is these individual characteristics that makes each client unique

Let's take a closer look at the structure of the skin.

There are two main layers of the skin:

The **epidermis** which is the outer thinner layer
The **dermis** which is the inner thicker layer

Below the dermis is the subcutaneous layer which attaches to underlying organs and tissues.

THE EPIDERMIS

The epidermis is the most superficial layer of the skin and consists of five layers of cells. The three outermost layers consist of dead cells as a result of the process of keratinisation; the cells in the very outermost layer are dead and scaly and are constantly being rubbed away by friction.

KEY NOTE

Keratin is the tough fibrous protein found in the epidermis, the hair and the nails. The keratin found in the skin is constantly being shed. Keratinisation refers to the process that cells undergo when they change from living cells with a nucleus (which is essential for growth and reproduction) to dead, horny cells without a nucleus. Cells which have undergone keratinisation are therefore dead. Keratin has a protective function in the skin as the keratinised cells form a waterproof covering, helping to stop the penetration of bacteria and protect the body from minor injury.

The inner two layers are composed of living cells. The epidermis does not have a system of blood vessels, only a few nerve endings which are present in the lower epidermis. Therefore, all nutrients pass to the cells in the epidermis from blood vessels in the deeper dermis. The five layers of cells of the epidermis are as follows:

- the basal cell layer
- the prickle cell layer
- the granular layer
- the clear layer
- the horny layer

BASAL CELL LAYER

This is the deepest of the five layers. It consists of a single layer of column cells on a basement membrane which separate the epidermis from the dermis. In this layer, the new epidermal cells are constantly being reproduced. These cells last about six weeks from reproduction or *mitosis* before being discarded into the horny layer. New cells are therefore formed by division, pushing adjacent cells towards the skin's surface. At intervals between the column cells, which divide to reproduce, are the large star-shaped cells called melanocytes, which form the pigment melanin, the skin's main colouring agent.

──────── KEY NOTE ────────

Melanin is produced by special cells called melanocytes, which are found in the basal cell layer of the epidermis. Melanocytes have finger-like projections which are capable of injecting melanin granules into neighbouring cells of the epidermis. This explains why melanin is found in the basal layer, prickle cell layer and granular layer of the epidermis. Melanin is responsible for the colour of the skin and hair and helps protect the deeper layers of the skin from the damaging effects of ultra violet radiation.

PRICKLE CELL LAYER

This is known as the prickle cell layer because each of the rounded cells contained within it have short projections which make contact with the neighbouring cells and give them a prickly appearance. The living cells of this layer are capable of dividing by the process mitosis.

GRANULAR LAYER

This layer consists of distinctly shaped cells, containing a number of granules which are involved in the hardening of the cells by the process keratinisation. This layer links the living cells of the epidermis to the dead cells above.

CLEAR LAYER

This layer consists of transparent cells which permit light to pass through. It consists of three or four rows of flat dead cells, which are completely filled with keratin; they have no nuclei as the cells have undergone mitosis. The clear layer is very shallow in facial skin but thick on the soles of the feet and palms of the hands and is generally absent in hairy skin.

HORNY LAYER

This is the most superficial, outer layer, consisting of dead, flattened, keratinised cells which have taken approximately a month to travel from the germinating layer. This outer layer of dead cells is continually being shed: this process is known as desquamation.

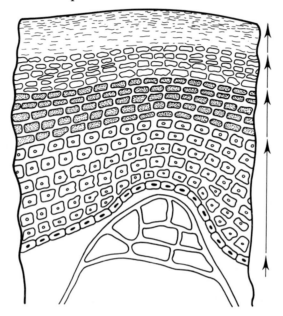

Figure 1
Cell regeneration of the epidermis

Cell regeneration occurs in the epidermis by the process mitosis. It takes approximately a month for a new cell to complete its journey from the basal cell layer where it is reproduced to the granular layer, where it becomes keratinised to the horny layer where it is desquamated.

THE DERMIS

The dermis is the deeper layer of the skin and its key functions are to provide support, strength and elasticity. The dermis has a superficial *papillary* layer and a deep *reticular* layer. The superficial papillary layer is made up of adipose connective tissue and is connected to the underside of the epidermis by cone-shaped projections called dermal papillae,which contain nerve endings and a network of blood and lymphatic capillaries. The many dermal papillae of the papillary layer form indentations in the overlying epidermis, giving it an irregular appearance.

─────────────── **KEY NOTE** ───────────────

The key function of the papillary layer of the dermis is to provide vital nourishment to the living layers of the epidermis above.

The deep reticular layer is formed of tough fibrous connective tissue which contains the following:

- *collagen* fibres containing the protein collagen, which gives the skin its strength and resilience
- *elastic* fibres containing a protein called elastin, which gives the skin its elasticity
- *reticular* fibres which help to support and hold all structures in place

These fibres all help maintain the skin's tone.

Cells present in the dermis are as follows:

- *mast* cells which secrete histamine, causing dilation of blood vessels to bring blood to the area. This occurs when the skin is damaged or during an allergic reaction
- *phagocytic* cells, which are white blood cells that are able to travel around the dermis destroying foreign matter and bacteria
- *fibroblasts*, which are cells that form new fibrous tissue

─────────────── **KEY NOTE** ───────────────

The principal function of the dermis is to provide nourishment to the epidermis and to give a supporting framework to the tissues.

BLOOD SUPPLY

Unlike the epidermis, the dermis has an abundant supply of blood vessels which run through the dermis and the subcutaneous layer of the skin. Arteries carry oxygenated blood to the skin via arterioles (small arteries) and these enter the dermis from below and branch into a network of capillaries around active or growing structures. These capillary networks form in the dermal papillae to provide the basal cell layer of the epidermis with food and oxygen. These networks also surround two appendages of the skin: the sweat glands and the erector pili muscles, which both have important functions in the skin. The capillary networks drain into venules, small veins which carry the deoxygenated blood away from the skin and remove waste products.

The dermis is therefore well supplied with capillary blood vessels to bring nutrients and oxygen to the germinating cells in the epidermis and to remove waste products from them.

LYMPH VESSELS

The lymphatic vessels are numerous in the dermis and generally accompany the course of veins. They form a network through the dermis, allowing removal of waste from the skin's tissues. Lymph vessels are found around the dermal papillae, glands and hair follicle.

NERVES

Nerves are widely distributed throughout the dermis. Most nerves in the skin are sensory nerves, which send signals to the brain and are sensitive to heat, cold, pain, pressure, touch. Branched nerve endings, which lie in the papillary layer and hair root, respond to touch and temperature changes. Nerve endings in the dermal papillae are sensitive to gentle pressure and those in the reticular layer are responsive to deep pressure.

The dermis also has motor nerve endings, which relay impulses from the brain and are responsible for the dilation and constriction of blood vessels, the secretion of perspiration from the sweat glands and the contraction of the arector pili muscles attached to hair follicles.

THE SUBCUTANEOUS LAYER

This is a thick layer of connective tissue found below the dermis. The tissues areolar and adipose are present in this layer to help support delicate structures such as blood vessels and nerve endings.

The subcutaneous layer contains the same collagen and elastin fibres as the dermis and also contains the major arteries and veins which supply the skin, forming a network throughout the dermis. The fat cells contained within this layer help to insulate the body by reducing heat loss. Below the subcutaneous layer of the skin lies the subdermal muscle layer.

———— K E Y N O T E ————

As therapeutic treatments involve the stimulation of the tissues of the skin, they have an effect on both the superficial epidermis and the deeper dermis. One of the common skin reactions to treatments given is the creation of an *erythema*. This occurs when the blood capillaries just below the epidermis dilate and a reddening appears on the skin's surface. This indicates that the deeper layers of the skin such as the dermal and subdermal muscle layers are being stimulated in reaction to treatment.

THE APPENDAGES OF THE SKIN

► TASK I

The skin has several appendages which are formed in the epidermis and extend down into the dermis. *With reference to your textbook, complete the following table. The first one has been completed for you.*

TABLE I: *Appendages of the skin*

Appendage	Structure	Location	Functional significance
Hair follicle	sac-like indentation in the epidermis	extends from epidermis (down) to dermis: found all over body except palms of hands and soles of feet.	provides hair with vital source of nourishment: contains dermal papilla which supplies blood to hair
Hair			
Erector pili muscle			
Sebaceous gland			
Sweat gland (eccrine)			
Sweat gland (apocrine)			

THE STRUCTURE OF THE SKIN

◆◆ TASK 2

Using the information you have researched so far during this chapter, label the following cross section diagram of the skin.

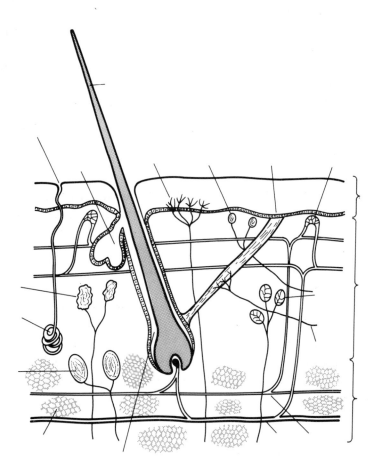

Figure 2
Cross section of skin

THE FUNCTIONS OF THE SKIN

The skin has several important functions:

PROTECTION

The skin acts as a protective organ in the following ways:

- the film of sebum and sweat on the surface of the skin, known as the acid mantle, acts as an anti-bacterial agent to help prevent the multiplication of micro-organisms on the skin
- the fat cells in the subcutaneous layer of the skin help protect bones and major organs from injury
- melanin, which is produced in the basal cell layer of the skin, helps to protect the body from the harmful effects of ultra violet radiation
- the cells in the horny layer of the skin overlap like scales to help prevent micro-organisms from penetrating the skin and to prevent excessive water loss from the body

TEMPERATURE REGULATION

The skin helps to regulate body temperature in the following ways:

- when the body is losing too much heat, the blood capillaries near the skin surface contract to keep warm blood away from the surface of the skin and closer to major organs
- the arector pili muscles raise the hairs and trap air next to the skin when heat needs to be retained
- the adipose tissue in the dermis and the subcutaneous layer helps to insulate the body against heat loss
- when the body is too warm, the blood capillaries dilate to allow warm blood to flow near to the surface of the skin in order to cool the body
- the evaporation of sweat from the surface of the skin will also assist in cooling the body

SENSITIVITY

The skin is very sensitive to various stimuli, due to its many sensory nerve endings which can detect changes in temperature and pressure, and register pain.

EXCRETION

The eccrine glands of the skin produce sweat, which helps to remove some waste materials from the skin such as urea, uric acid, ammonia, and lactic acid.

SKIN DISEASES AND DISORDERS

A therapist needs to be able to recognise diseases and disorders of the skin to enable proper judgement to be made as to whether a client may be treated in a particular area or whether they need to be referred to their medical practitioner for treatment and advice.

A therapist must therefore be particularly vigilant in recognising infectious skin diseases and disorders to prevent cross infection and to maintain a hygienic environment for themselves, their colleagues and clients.

⊶ TASK 3

Using a recommended textbook, complete the following table by identifying the skin disease or disorder and whether it is treatable or non-treatable therapeutically.

TABLE 2: *Skin diseases and disorders*

Disorder/disease	Appearance	Cause	Treatable/ non-treatable
	oily skin, papules, pustules, cysts, scars	overproduction of sebum from sebaceous glands; puberty/hormonal	
	flushing of cheeks and nose: dilated capillaries; progresses to papules, pustules, scales	unknown, factors such as spicy food, alcohol, heat, menopause, emotional stress may be contributory	
	skin appears greasy, comedones develop; may develop into acne vulgaris	increase in sebum from sebaceous glands/puberty	
	begins as small red nodule, which forms pocket of bacteria around base of hair follicle	bacterial infection	

	weeping blisters which form honey-coloured crusts	bacterial infection	
	area of redness which blisters and forms a crust around lips on face	viral infection	
	itchy redness; blisters develop along path of sensory nerves; affects chest, back, face	viral infection	
	small red papules which increase in size to form rings which are itchy, inflamed and scaly	fungal infection	
	areas of flaking skin between the toes; skin becomes soft and wet, skin on sole of foot may be cracked	fungal infection	
	severe, itchy, scaly skin	parasitic condition: female mite burrows and lays eggs into horny layer of skin	
	patches of skin redness which appear cracked and scaly: skin may weep	genetic, external/ internal influences	
	inflammation of skin – skin appears red and dry	contact with external irritants	
	red plaques of skin covered by white, silvery scales; affects face,	faulty keratinisation of skin; stress can be aggravating factor	

	elbow, knees, chest, abdomen, nail		
	found on face, forehead, back of hands, front of knees – smooth in texture with flat top	viral infection	
	found on soles of feet; pea sized	viral infection	
	irregular areas of increased pigmentation, usually found on face	occurs during pregnancy, sometimes when taking contraceptive pill, due to stimulation of melanin by female hormone oestrogen	
	areas of skin which have a lack of pigmentation	unknown – basal cell layer of epidermis no longer produces melanin	
	inner eyelid and eyeball appear red and sore	bacterial infection following irritation of conjunctiva of eye	

Note: *There are other skin diseases and disorders which may be researched with further study.*

? SELF-ASSESSMENT QUESTIONS

1. List the five layers of the epidermis

....

....

....

....

....

2. *Briefly* describe the process of cell regeneration in the epidermis

....

....

....

....

....

....

....

3. Name the protein found in the epidermis. What is its function?

....

....

....

....

4. What is melanin and where is it found in the skin?

....

....

....

5. What type of tissue does the dermis contain?

....

6. Where is muscle located in relation to the layers of the skin?

....

....

7. What is the function of the dermis?

....

....

8. List the sensations the sensory nerves of the skin enable us to feel

....

....

....

9. What is the function of the subcutaneous layer of the skin?

....

....

....

10 i) List two ways in which the skin acts as a protective organ

....

....

ii) List two ways in which the skin helps to regulate body temperature

....

....

iii) List two other functions of the skin

....

....

....

Chapter 2

————————〜————————

THE HAIR

In order to be successful in removing hair temporarily or permanently it is imperative for a therapist to have a working knowledge of the anatomical structure of a hair.

By the end of this chapter you will be able to relate the following to your practical work carried out in the salon:

- the structural and functional parts of a hair
- the growth cycle of a hair
- the different types of hair growth on the body

Hair is an important appendage of the skin which grows from a sac-like depression in the epidermis called a hair follicle. Hair grows all over the body, with the exception of the palms of the hands and the soles of the feet and is a sexual characteristic.

The primary functions of hair are:

PHYSICAL PROTECTION
- the eyelashes act as a line of defence by preventing the entry of foreign particles into the eyes and helping shade the eyes from the sun's rays
- eyebrow hairs help to divert water and other chemical substances away from the eyes
- hairs lining the ears and the nose trap dust and help to prevent bacteria from entering the body
- body hair acts as a protective barrier against the sun and helps to protect us against the cold with the help of the erector pili muscle

PREVENTING FRICTION
- body hair is present on the body where muscular action causes friction

THE STRUCTURE OF A HAIR

The hair is composed mainly of the protein keratin and therefore is a dead structure.

Longitudinally the hair is divided into three parts:

- hair *shaft*: the part of the hair which lies above the surface of the skin
- hair *root*: the part which is found below the surface of skin
- hair *bulb*: the enlarged part at the base of the hair root

Internally the hair has three layers:

- **cuticle**: the outer layer, made up of transparent protective scales which overlap one another. The cuticle protects the cortex and gives the hair its elasticity
- **cortex**: the middle layer, made up of tightly packed keratinised cells containing the pigment melanin which gives the hair its colour. The cortex helps to give strength to the hair
- **medulla**: the inner layer, made up of loosely connected keratinised cells and tiny air spaces. This layer of the hair determines the sheen and colour of hair due to the reflection of light through the air spaces

☞ TASK I

With reference to your recommended textbook, complete the following table to show the location and function of the individual parts of a hair structure.

TABLE 3: *Structure of a hair*

Structure	Location	Functional significance
Connective tissue sheath		
Outer root sheath		
Inner root sheath		

TABLE 3 *continued*

Structure	Location	Functional significance
Bulb		
Matrix		
Dermal papilla		

THE HAIR IN ITS FOLLICLE

►○ TASK 2

Based on the information you have researched, label the following cross section diagram of the hair in its follicle.

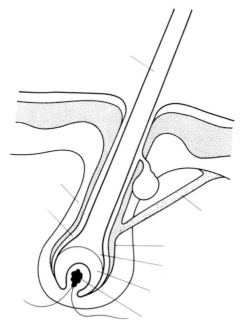

Figure 3
Hair in its follicle

FACTORS ABOUT HAIR GROWTH

- hair begins to form in the foetus from the third month of pregnancy
- growth of hair originates from the matrix, which is the active growing area where cells divide and reproduce by mitosis
- living cells, which are produced in the matrix, are pushed upwards away from their source of nutrition, they die and are converted to keratin to produce a hair
- hair has a growth pattern which ranges from approximately four to five months for an eyelash hair to approximately four to seven years for a scalp hair
- hair growth is affected by illness, diet and hormonal influences

THE GROWTH CYCLE OF A HAIR

Each hair has its own growth cycle and undergoes three distinct stages of development.

ANAGEN

- active growing stage
- lasts from a few months to several years
- hair germ cells reproduce at matrix
- new follicle is produced which extends in depth and width
- the hair cells pass upwards to form hair bulb
- hair cells continue rising up the follicle and as they pass through the bulb they differentiate to form individual structures of hair
- inner root sheath grows up with the hair, anchoring it into the follicle
- when cells reach upper part of bulb they become keratinised
- two-thirds of way up follicle, hair leaves inner root sheath and emerges onto surface of skin

---

KEY NOTE

An anagen hair receives its nourishment to grow from dermal papilla and when removed has a visibly developed bulb and inner root sheath intact.

CATAGEN

- lasts approximately two to four weeks
- transitional stage – from active to resting

- hair separates from dermal papilla and moves slowly up the follicle
- follicle below retreating hair shrinks
- hair rises to just below level of sebaceous gland where the inner root sheath dissolves and the hair can be brushed out

KEY NOTE

In a catagen hair a column of epithelial cells remains in contact with the dermal papilla and as the hair breaks away from the bulb it receives its nourishment from the follicle wall. A catagen will be have no bulb visible when removed and will appear shorter and dehydrated.

TELOGEN

- short resting stage
- shortened follicle rests until stimulated again
- hair is shed onto skin's surface
- new replacement hair begins to grow

KEY NOTE

A telogen hair has a diminished blood supply and when removed has small brim-like fibres at its end.

THE STAGES OF HAIR GROWTH

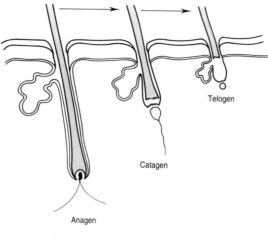

Figure 4
The hair growth cycle

———————— **KEY NOTE** ————————

Each follicle has a life expectancy of its own and once a follicle has produced its expected number of hairs it will no longer function.

DIFFERENT TYPES OF HAIR GROWTH

As there is a continuous cycle of hair growth the amount of hair on the body remains fairly constant. However, hair growth will vary from client to client and from area to area. A new client coming to the salon for a hair removal treatment should be made aware of the fact that hair growth occurs in three stages, which will result in the hair being at different lengths both above the skin and below it.

———————— **KEY NOTE** ————————
– INGROWING HAIRS

Ingrowing hairs present a problem when carrying out hair removal treatments as the hair grows back on itself due to the follicle becoming blocked. This may be because the hair has grown so weak that it can no longer push up through the skin so it grows parallel to the skin.

SKIN REACTION TO HAIR REMOVAL TREATMENTS

Whilst carrying out hair removal treatments it is important to remember that the hair follicle is part of the skin's structure, therefore any treatment which affects the hair is also going to affect the skin.

Once a hair has been removed the maximum amount of blood will be sent straight to the area being treated to heal and protect the skin. This is a normal reaction of the skin and extra blood that has been sent to the treated area will soon be diverted again within a few hours of treatment. As the treated area of skin will have open follicles, it is vital that a client adheres strictly to after-care advice specified as open follicles offer bacteria an easy entry into the body.

? SELF-ASSESSMENT QUESTIONS

1. Name the three layers of the hair

   ....

   ....

   ....

2. Where are hair cells produced?

   ....

3. Where does hair growth originate from?

   ....

4. Which structure supplies the hair follicle with nerves and blood?

   ....

   ....

5. Which structure grows upwards with the hair and what is its function?

   ....

   ....

   ....

6. Name two appendages of the skin associated with the hair

   ....

   ....

7. What are the stages of hair growth?

   ....

   ....

   ....

8. How would you recognise a growing hair after it has been removed?

....

....

Chapter 3

THE NAIL

A sound knowledge of the structure of the nail and its functional parts is fundamental to understanding how a nail grows.

A competent therapist needs to be able to:

- understand the process of nail growth to explain the benefits of salon treatments and the consequences any damage may have
- correctly identify the functional parts of the nail to apply treatments correctly and avoid any resulting damage

By the end of this chapter, you will be able to relate the following knowledge to your practical work carried out in the salon:

- the structure of the nail in its bed
- the functional significance of the individual parts of the nail
- the process of nail growth
- adverse treatable nail conditions and non-treatable nail diseases

The nail is an important appendage of the skin and is an extension of the clear layer of the epidermis. It is composed of horny flattened cells which undergo a process of keratinisation, giving the nail a hard appearance.

It is the protein keratin which helps to make the nail a strong but flexible structure. The part of the nail the eye can see is dead as it has no direct supply of blood, lymph and nerves, and all nutrients are supplied to the nail via the dermis.

FUNCTIONS OF THE NAIL

The nail has several important functions:

- it forms a protective covering at the ends of the phalangeal joints of the fingers and the toes, helping to support the delicate network of blood vessels and nerves at the end of the fingers

- it is a useful tool in that it enables us to concentrate touch to manipulate small objects and to scratch surfaces

THE STRUCTURE OF THE NAIL

The nail has several important anatomical regions (indicated in Table 4).

⊶ TASK I

With reference to your recommended textbook, complete the following table to show the location and functional significance of the individual parts of the nail structure. The first one has been completed for you.

TABLE 4: *Nail structure*

Part of nail	Location	Functional significance
Matrix	immediately below cuticle	most important feature of nail: site of nail growth, where living cells are produced; receives rich supply of blood and lymph; area from which health of nail is determined
Nail bed		
Cuticle		

TABLE 4: *continued*

Part of nail	Location	Functional significance
Lanula		
Nail plate		
Nail wall		
Nail grooves		
Free edge		

CROSS SECTION OF THE NAIL

TASK 2

Based on the information you have researched, label the above structures on the following cross section diagram of the nail.

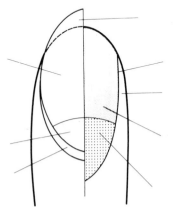

Figure 5
Cross section of the nail in its bed

FACTORS ABOUT NAIL GROWTH

- nails start growing on a foetus before the fourth month of pregnancy
- nail growth occurs from the nail matrix by cell division
- as new cells are produced in the matrix, older cells are pushed forward and are hardened by the process keratinisation, which forms the hardened nail plate
- as the nail grows it moves along the nail furrows at the sides of the nail, which helps to direct the nail growth along the nail bed
- it takes approximately six months for cells to travel from the lanula to the free edge of the nail
- the growth of a nail does not follow a growth cycle and hence growth is continuous throughout life
- the average growth rate of a nail is approximately 3 mm per month
- the growth rate of nails will vary from person to person and from finger to finger, with the index finger generally being the fastest to grow
- toe nails have a slower rate of growth than fingernails
- the rate of growth of a nail is faster in the summer due to an increase in cell division as a result of exposure to ultra violet radiation.

- a good blood supply is essential to nail growth; oxygen and nutrients are fed to the living cells of the nail matrix and nail bed
- protein and calcium are good sources of nourishment for the nails

Nail growth may be affected by the following factors:

- **ill health**: during illness your body will receive a reduced blood supply to the nail as it attempts to restore the rest of the body to good health
- **diet**: a nutritional deficiency can result in a diminished blood supply to the nail
- **age**: during ageing the growth of a nail slows down due to the fact that the blood vessels supplying the matrix and the nail become less efficient
- **poor technique**: if a heavy pressure is used when using manicure implements such as a cuticle knife, damage may be caused to the matrix cells resulting in ridges to the nail. This may only be temporary, as new cells produced in the matrix will replace the damaged ones, and depending on the extent of the damage the ridges may eventually grow out
- **an accident**: such as shutting a finger in the door. This may result in bruising and bleeding of the nail or even the complete removal of a nail. It could result in permanent malformation of the nail if the nail bed has become damaged

---

——————— K E Y N O T E ———————

It is essential when carrying out treatments for the hands and feet that care is taken to avoid damaging the nail matrix and nail bed as they are the fundamental parts of the nail concerned with growth.

COMMON NAIL DISEASES

☞ TASK 3

A therapist needs to be able to recognise the sign of disease in a nail.
With reference to your recommended textbook, research the following nail diseases and complete the following table.

TABLE 5: *Nail diseases*

Nail disease	Appearance	Cause	Treatable/ non-treatable
	inflammation of skin round nail, may be swollen, pus may be present	bacterial infection following damage to skin, e.g. using unsterilised manicure implements	non-treatable until infection subsides and skin healed: GP referral if chronic
	yellow discoloration at free edge which may spread down to nail root: nail plate appears spongy and furrowed	fungus which attacks nail bed and nail plate	non-treatable: GP referral

Note: *There are other nail diseases which may be researched with further study.*

SKIN DISORDERS AFFECTING HANDS AND FEET

TASK 4

A therapist needs to have a good knowledge of skin disorders affecting the hands and feet in order to decide whether therapeutic treatments are permissible or not.

With reference to your recommended textbook, identify the name of the disease or disorder and state whether it is treatable or non-treatable.

TABLE 6: *Skin disorders of hands and feet*

Disorder	Appearance	Cause	Type of disorder	Treatable/ non-treatable
	inflamed red skin: may be blistered, cracked and scaly	genetic external: internal irritants	inflammatory	

TABLE 6: *continued*

Disorder	Appearance	Cause	Type of disorder	Treatable/ non-treatable
	inflamed, red dry skin	contact with external irritants (e.g. detergents, lanolin, nickel, acids, alkalis)	inflammatory	
	red plaques, covered by white or silvery scales; usually affects knees, elbows, face, nails: when affecting nails, nail bed appears pitted	stress is contributory factor	chronic inflammatory	
	build up of pus on soft pad at tip of finger; thumb appears hot, swollen and inflamed	bacteria infecting nail fold	bacterial infection	
	skin-coloured/ as brownish raised surface on back of hands, or in association with nails. Varies in size from pinhead to size of pea	virus entering epidermis	viral infection	
	flaking of skin between toes; skin appears soft and wet: skin may be split; soles of feet may also be affected	fungus, contracted in damp, moist conditions	fungal infection	

? SELF-ASSESSMENT QUESTIONS

1. What are the functions of the nail?

....

....

....

....

2. What substance is the nail mainly composed of?

....

....

3. Which is the living part of the nail?

....

....

4. What is the function of the cuticle?

....

....

5. Which is the area of the nail where the cells start to harden?

....

6. What is the function of the nail grooves?

....

....

7. *Briefly* describe how a nail grows

....

....

....

....

....

....

Chapter 4

———————⌇———————

THE SKELETAL SYSTEM

The skeleton is made up of no fewer than 206 individual bones, which collectively form a strong framework for the body. Bones are like *landmarks* in the body and by tracing their outlines you can be accurate in describing the position of muscles, glands and organs in relation to your work as a therapist. Learning the positions of the bones of the skeleton is essential for learning the position of the superficial muscles, as bones must have muscle attachments to enable them to move.

At the back of this chapter you will find a glossary of anatomical terms to assist you in learning the precise positions of bones.

A competent therapist needs to be able to:

- have a good knowledge of the framework of the body in order to be able to apply treatments safely and effectively

By the end of this chapter you will be able to relate the following knowledge to your practical work carried out in the salon:

- The functions of the skeleton
- The names and positions of the primary bones of the skeleton
- Anatomical terms used to describe the position of the bones of the skeleton

FUNCTIONS OF THE SKELETON

The skeletal system is made up of all types of bones which form the skeleton or bony framework of the body. Before learning the individual bones of the skeleton, it is important to understand the functions of the skeleton as a whole.

SUPPORT

The skeleton bears the weight of all other tissues; without it we would be unable to stand up. Consider the bones of the vertebral column, the pelvis, the feet and the legs which all support the weight of the body.

SHAPE

The bones of the skeleton give shape to structures such as the skull, thorax and limbs; without soft tissue covering these areas we would look prehistoric!

PROTECTION OF VITAL ORGANS AND DELICATE TISSUE

The skeleton surrounds vital organs and tissue with a tough, resilient covering. For example the rib cage protects the heart and lungs and the vertebral column protects the spinal cord.

ATTACHMENT FOR MUSCLES AND TENDONS

To allow movements. Bones are like anchors which allow the muscle to function efficiently.

MOVEMENT

This happens as a result of co-ordinated action of muscles upon bones and joints. Bones are therefore levers for muscles.

FORMATION OF BLOOD CELLS

These develop in red bone marrow

MINERAL RESERVOIR

The skeleton acts as a storage depot for important minerals such as calcium, which can be released when needed for essential metabolic processes such as muscle contraction and conduction of nerve impulses.

The skeletal system is divided into two parts. The Axial Skeleton forms the main axis or central core of the body and consists of the following parts:

- The skull
- The vertebral column
- The sternum
- The ribs

The Appendicular Skeleton supports the appendages or limbs and gives them attachment to the rest of the body. It consists of the following parts:

- The shoulder girdle
- Bones of the upper limbs
- Bones of the lower limbs
- Bones of the pelvic girdle

THE AXIAL SKELETON

THE SKULL

The skull rests upon the upper end of the vertebral column and its bony structure can be described in two different parts:

- The cranium
- The face

The cranium is made up of eight flat irregular bones which provide the bony structure of the head and provide attachment for muscles.

The following bones of the skull are of relevance to a therapist as they form the major part of the skull:

- One **frontal**, which forms the anterior part of the roof of the skull, the forehead and the upper part of the eye sockets
- Two **parietal**, forming the upper sides of the skull and the back of the roof of the skull
- Two **temporal**, forming the sides of the skull below the parietal and above and around the ears
- One **occipital**, forming the back of the skull

MAJOR BONES OF THE SKULL

◖◆ TASK I

Based on the information given, label the major bones of the skull on the following diagram.

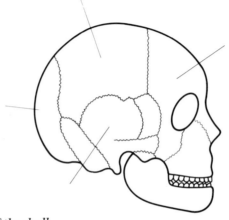

Figure 6
Major bones of the skull

THE FACE

There are fourteen facial bones in total, *seven of which are of relevance to a therapist as they contribute to the facial contours and provide attachment for various muscles of the face and neck*:

- Two **zygomatic** or cheek bones. *These are the most prominent of the facial bones*
- Two **maxilla**, which form the upper jaw and carry the upper teeth. *These are the largest bones of the face*
- One **mandible**, which forms the lower jaw. *It is important to note that this bone is the only movable bone of the skull as the rest are fused together by immovable joints*

MAJOR BONES OF THE FACE

✦ TASK 2

Based on the information given above label the major bones of the face on the following diagram.

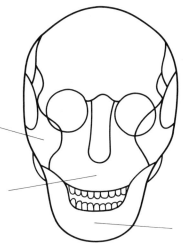

Figure 7
Major bones of the face

THE VERTEBRAL COLUMN

The vertebral column lies on the posterior of the skeleton, extending from the skull to the pelvis, providing a central axis to the body. It consists of 33 individual irregular bones called vertebrae; however, the bones of the base of the vertebral column, the sacrum and coccyx, are fused to give 24 movable bones in all.

The vertebral column is made up of the following:

- Seven cervical vertebrae

These are the vertebrae of the neck; the top two vertebrae, C1 the atlas, and C2 the axis are part of a pivot joint, which allow the head and neck to move freely. *These are the smallest vertebrae in the vertebral column.*

- Twelve thoracic vertebrae

These are the vertebrae of the mid spine and they lie in the thorax, where they articulate with the ribs. *These vertebrae lie flatter and downwards to allow for muscular attachment of the large muscle groups on the back. They can be easily felt as you run your fingers down the spine.*

- Five lumbar vertebrae

These lie in the lower back and are much larger in size than the vertebrae above them as they are designed to support more body weight. *These vertebrae can be easily felt on the lower back due to their large shape and width.*

- Five sacral vertebrae

This is a very flat triangular shaped bone and lies in between the pelvic bones. It is made up of five bones which are fused together. *A characteristic feature of the sacrum is the sacral holes, of which there are eight. It is through these holes that nerves and blood vessels penetrate.*

- Four coccygeal vertebrae

These are made up of four bones which are fused together and are sometimes referred to as the tail bone.

— KEY NOTE —

In between the vertebrae lie a padding of fibrocartilage called the intervertebral discs. These give the vertebrae a certain degree of flexibility and also act as shock absorbers in between the vertebrae, cushioning any mechanical stress that may be placed upon them.

FUNCTIONS OF THE VERTEBRAL COLUMN

Now we have considered the individual structure of the vertebrae, let us consider the functions of the vertebral column as a whole:

- The vertebral column provides a strong and slightly flexible axis to the skeleton
- By way of its different shaped vertebrae with their roughened surfaces, it is able to provide a surface for the attachment of muscle groups
- The vertebral column also has a protective function as it protects the delicate nerve pathways of the spinal cord

Therefore, the vertebral column mirrors the primary functions of the skeleton in its supportive and protective roles.

THE THORAX

This is the area of the body enclosed by the ribs, which provides protection for the heart and lungs. Essential organs contained within this cavity include:

- The sternum
- The ribs
- 12 thoracic vertebrae

THE STERNUM

This is commonly referred to as the breast bone and is a flat bone lying just beneath the skin in the centre of the chest. The sternum is divided into three parts:

- The manubrium, which is the top section
- The main body, which is the middle section
- The xiphoid process, which is the bottom section

The top section of the sternum articulates with the clavicle and the first rib. The middle section articulates with the costal cartilages which link the ribs to the sternum. The bottom section provides a point of attachment for the muscles of the diaphragm and the abdominal wall.

THE RIBS

There are 12 pairs of ribs. They articulate with the thoracic vertebrae, posteriorly. Anteriorly, the first 10 pairs attach to the sternum via the costal cartilages, the first 7 directly (known as the true ribs), the remaining 3 indirectly (known as the false ribs). The last 2 ribs have no anterior attachment and are called the false ribs.

THE APPENDICULAR SKELETON

This consists of the shoulder girdle, the bones of the upper and lower limbs and the pelvic girdle.

The **shoulder girdle** connects the upper limbs with the thorax and consists of four bones:

- Two clavicle
- Two scapula

The **scapula** is a large flat bone, triangular in outline, which forms the posterior part of the shoulder girdle. The scapula articulates with the clavicle and the humerus and serves as a point of muscle attachment which connects the shoulder girdle with the trunk and upper limbs.

The **clavicle** is a long slender bone with a double curve. It forms the anterior portion of the shoulder girdle. At its medial end it articulates with the top part of the sternum and at its lateral end it articulates with the scapula.

The clavicle provides the only bony link between the shoulder girdle and the axial skeleton. The arrangement of bones and the muscle attached to the scapula and the clavicle allow for a considerable amount of movement of the shoulder and the upper limbs.

BONES OF THE NECK AND SHOULDER

☞ TASK 3

Label the major bones of the neck and the shoulder girdle on the following diagram.

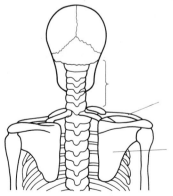

Figure 8
Bones of the neck and shoulder

The **upper limb** consists of the following bones:

- Humerus
- Radius
- Ulna
- Carpals
- Metacarpals
- Phalanges

The **humerus** is the long bone of the upper arm. The head of the humerus articulates with the scapula, forming the shoulder joint. The distal end of the bone articulates with the radius and ulna to form the elbow joint.

The **ulna** and **radius** are the long bones of the forearm. The two bones are bound together by a fibrous ring, which allows a rotating movement in which the bones pass over each other. The ulna is the bone of the little finger side and is the longer of the two forearm bones. The radius is situated on the thumb side of the forearm and is shorter than the ulna. The joint between the ulna and the radius permits a movement called pronation. This is when the radius moves obliquely across the ulna so that the thumb side of the hand is closest to the body.

The movement called supination takes the thumb side of the hand to the lateral side. The radius and the ulna articulate with the humerus at the elbow and the carpal bones at the wrist.

THE WRIST AND HAND

The wrist consists of eight small bones of irregular size which are collectively called carpals. They fit closely together and are held in place by ligaments. The carpals are arranged in two groups of four; those of the upper row articulate with the ulna and the radius and the lower row articulate with the metacarpals. There are five long metacarpal bones in the palm of the hand; their proximal ends articulate with the wrist bones and the distal ends articulate with the finger bones. There are fourteen phalanges, which are the finger bones, two of which are in the thumb or pollex and three in each of the other digits.

The **lower limb** consists of the following bones:

- Femur
- Tibia
- Fibula
- Patella

- Tarsals
- Metatarsals
- Phalanges

The **femur** is the longest bone in the body and has a shaft and two swellings at each end. The proximal swelling has a rounded head like a ball which fits into the socket of the pelvis to form the hip joint. Below the neck are swellings called trochanters which are sites for muscle attachment. The distal ends of the femur articulate with the patella or knee cap.

The **tibia** and **fibula** are long bones of the lower leg. The tibia is situated on the anterior and medial side of the lower leg. It has a large head, where it joins the knee joint and the shaft leads down where it forms part of the ankle. The tibia is the larger of the two bones of the lower leg and thus carries the weight of the body. The fibula is situated on the lateral side of the tibia in the lower leg and is the shorter and thinner of the two bones. The end of the fibula forms part of the ankle on the lateral side.

THE FOOT

There are seven bones in the foot which are collectively called the tarsals. Of particular relevance to a therapist is the talus bone, which articulates with the tibia and the fibula and bears the weight of the legs and the calcaneum, and is an important site of attachment for muscles of the calf.

There are five metatarsals forming the dorsal surface of the foot. Fourteen phalanges form the toes, two of which are in the hallux or big toe and three to each of the other digits.

ARCHES OF THE FOOT

The bones of the feet form arches which are designed to support body weight and to provide leverage when walking.

There are four arches of the foot:

Medial longitudinal arch, which runs along the medial side of foot from the calcaneum bone to the end of the metatarsals
Lateral longitudinal arch, which runs along the lateral side of the foot from the calcaneum bone to the end of the metatarsals
Anterior transverse arch, which run across the top of the foot, across the lower end of the metatarsals
Posterior transverse arch, which runs across the bottom of the foot, across the lower end of the metatarsals

THE PELVIC GIRDLE

The pelvic girdle consists of two hip bones which are joined together at the back by the sacrum and at the front by the symphysis pubis. Each hip bone consists of three separate bones which are fused together:

- The ilium
- The ischium
- The pubis

The ilium is the largest of the three bones and its upper border is called the iliac crest, which is an important site of attachment for muscles of the anterior and posterior abdominal walls.

The ischium is the bone at the lower end and forms the posterior part of the pelvic girdle.

The pubis bone is situated on the anterior part of the pelvis and it is via the symphysis pubis that the two pubis bones are linked.

FUNCTIONS OF THE PELVIC GIRDLE

Like the vertebral column, the pelvic girdle mirrors the primary functions of the skeleton, insofar as it has a role in *supporting* the vertebral column and the body's weight and that it offers *protection* by encasing delicate organs such as the uterus and bladder.

PRIMARY BONES OF THE SKELETON

⊷ TASK 4

Label the primary bones of the skeleton on the following diagrams.

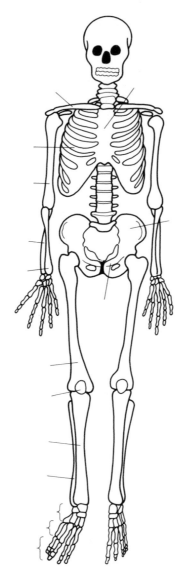

Figure 9
Anterior view of primary bones of the skeleton

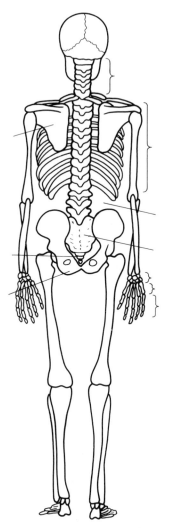

Figure 10
Posterior view of primary bones of the skeleton

GLOSSARY OF ANATOMICAL TERMS

An anatomical position is determined from a central imaginary line running down the centre of the body. In order to give precise descriptions of the positions of certain bones and muscles, it is important to use anatomical terminology. Using anatomical terms is like learning a new language. It will enable you to state precisely where a bone or muscle is situated in the body:

Anterior: front surface of the body

Posterior: back surface of the body

Lateral: away from the midline

Medial: towards the midline

Superior: upper surface of a structure, towards the head

Inferior: below or lower surface of a structure, away from the head

Proximal: nearest to the midline or point of origin of a part

Distal: furthest away from the midline or point of origin of a part

Dorsal: top of the foot is the dorsal surface

Plantar: sole of the foot is the plantar surface

? SELF-ASSESSMENT QUESTIONS

1. List the functions of the skeleton

....

....

....

....

....

....

....

....

....

2. Name the major bones of the skull

....

....

3. What are the functions of the vertebral column?

....

....

....

....

4. Name the bones which form the shoulder girdle

....

....

5. Which bone of the upper limb articulates with the scapula?

....

6. Name the bones which collectively form the wrist; how many are there?

....

7. Name the bones of the lower leg

....

8. Name the bone of the lower limb which fits into the socket of the pelvis

....

....

9. Name and give the positions of the bones of the pelvis

....

....

....

10. What are the functions of the pelvis?

....

....

....

11. Explain the following anatomical terms
    i) anterior

....

    ii) lateral

....

    iii) proximal

....

Chapter 5

JOINTS

We have so far seen that the skeleton comprises bones which provide support and protection for our body. Bones must, however, be *linked* together in order to facilitate their supportive role and to allow movement to occur. It is *joints* that provide the *link* between bones of the skeletal system. At the end of this chapter you will find a glossary of terms to assist you in learning the angular movements that occur at joints.

A competent therapist needs to be able to:

* have a good knowledge of joints to understand how body movements occur

By the end of this chapter you will be able to relate the following knowledge to your practical work carried out in the salon:

* the types of joints and their range of movement
* the structure of a synovial joint
* anatomical terms used to describe the range of movements

A joint is formed where two or more bones or cartilage meet and is otherwise known as an articulation. Where a bone is a *lever* in a movement, the joint is the *fulcrum* or the support which steadies the movement and allows the bone to move in certain directions.

Joints are classified according to the *degree* of movement possible at each one. There are three main joint classifications:

* Fibrous, where no movement is possible. Also known as fixed joints
* Cartilaginous, where slight movement is possible
* Synovial, which are freely movable joints

FIBROUS JOINTS

These are immovable joints, which have tough fibrous tissue between the bones. Often the edges of the bones are dovetailed together into one another, as in the sutures of the skull. Some examples of fibrous joints include the joints between the teeth and between the maxilla and mandible of the jaw.

CARTILAGINOUS JOINTS

These are slightly movable joints, which have a pad of fibrocartilage between the end of the bones making the joint. The pad acts as a shock absorber. Some examples of cartilaginous joints are those between the vertebrae of the spine and at the symphysis pubis, in between the pubis bones.

SYNOVIAL JOINTS

These are freely movable joints which have a more complex structure than the fibrous or cartilaginous joints. *Before looking at the different types of synovial joints it is important to have an understanding of the general structure of a synovial joint.*

THE GENERAL STRUCTURE OF A SYNOVIAL JOINT

A synovial joint has a space between the articulating bones which is known as the synovial cavity. The surface of the articulating bones is covered by hyaline cartilage, which is supportive to the joint by providing a hard wearing surface for the bones to move against one another with the minimum of friction. The synovial cavity and the cartilage are encased within a fibrous capsule, which helps to hold the bones together to enclose the joint. This joint capsule is reinforced by tough sheets of connective tissue called ligaments, which bind the articular ends of bones together.

The joint capsule is reinforced enough to allow strength to resist dislocation but is flexible enough to allow movement at the joint. The inner layer of the joint capsule is formed by the synovial membrane which secretes a sticky oily fluid called synovial fluid which lubricates the joint and nourishes the hyaline cartilage. As the hyaline cartilage does not have a direct blood supply it relies on the synovial fluid to deliver its oxygen and nutrients and to remove waste from the joint, which is achieved via the synovial membrane.

THE STRUCTURE OF A SYNOVIAL JOINT

TASK 1

Label the important functional parts of a synovial joint.

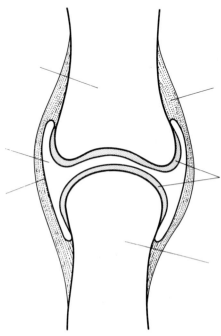

Figure 11
Structure of a synovial joint

Synovial joints are classified into six different types according to their shape and the movements possible at each one. The degree of movement possible at each synovial joint is dependent on the type of synovial joint and its articulations.

TYPES OF SYNOVIAL JOINTS

TASK 2

Using your recommended textbook, complete the following:

i) identify the type of synovial joint
ii) identify the movements possible at each joint
iii) give two examples of where the joint is found in the body

1. ....

This type of joint is where the rounded head of one bone fits into a cup shaped cavity of another bone. Movements possible at this joint are

....

....

Give two examples of this type of joint

....

....

2. ....

This type of joint is where the rounded surface of one bone fits the hollow surface of another. Movement is only possible in one plane with this type of joint. Movements possible at this joint are

....

....

Give two examples of this type of joint

....

....

3. ....

This type of joint occurs where two flat surfaces of bone slide against one another. Movements possible at this joint are

....

....

Give two examples of this type of joint

....

....

4. ....

This type of joint occurs where a process of bone rotates in a socket. Movements possible at this joint are

....

....

Give two examples of this type of joint

....

....

5. ....

This type of joint occurs when a rounded surface fits a hollow surface of such form that movement is possible in two *planes.* Movements possible at this joint are

....

....

Give two examples of this type of joint

....

....

6. ....

This type of joint is shaped like a saddle and its articulating surfaces of bone have both rounded and hollow surfaces so that the surface of one bone fits the complementary surface of the other. Movements possible at this joint are

....

....

Give two examples of this type of joint

....

....

GLOSSARY OF ANGULAR MOVEMENTS POSSIBLE AT JOINTS

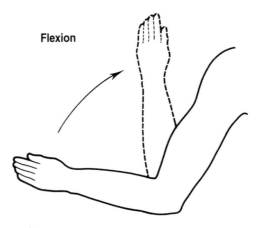

Figure 12
Flexion

Flexion: bending of a body part at a joint so that the angle between the bones is decreased

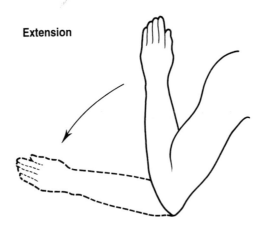

Figure 13
Extension

Extension: straightening of a body part at a joint so that the angle between the bones is increased

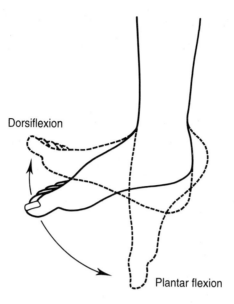

Figure 14
Dorsiflexion/plantar flexion

Dorsiflexion: upward movement of the foot so that feet point upwards
Plantar flexion: downward movement of the foot so that feet face downwards
towards the ground

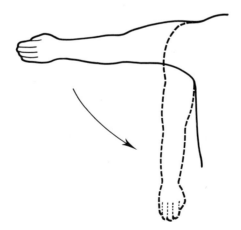

Figure 15
Adduction

Adduction: movement of a limb towards the midline

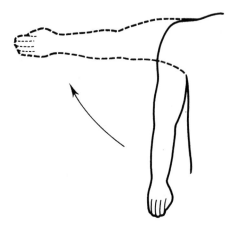

Figure 16
Abduction

Abduction: movement of a limb away from the midline

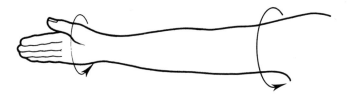

Figure 17
Rotation

Rotation: movement of a bone around an axis (180 degrees)

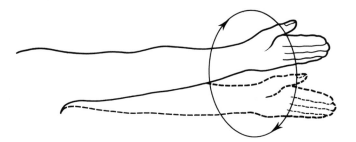

Figure 18
Circumduction

Circumduction: a circular movement of a joint (360 degrees)

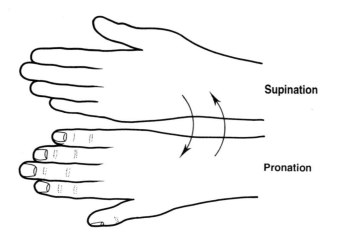
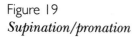

Figure 19
Supination/pronation

Supination: turning the hand so that the palm is facing upwards
Pronation: turning the hand so that the palm is facing downwards

Figure 20
Eversion

Eversion: soles of the feet face outwards

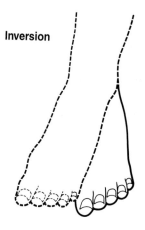

Figure 21
Inversion

Inversion: soles of the feet face inwards

❓ SELF-ASSESSMENT QUESTIONS

1. Name the three main joint classifications

....

....

2. i) State the joint classifications for the following:
   e.g. the hip is classified as a synovial joint but is also a ball-and-socket joint.
   a) joints between the skull bones

....

   b) the atlas and axis

....

   c) the symphysis pubis

....

   d) the wrist joint

....

e) the knee joint

....

f) the trapezium and the metacarpal of the thumb

....

3. Describe the following terms
   a) flexion

....

b) abduction

....

c) rotation

....

d) extension

....

e) supination

....

Chapter 6

THE MUSCULAR SYSTEM

The muscular system comprises over six hundred individual specialised cells called muscles which are primarily concerned with movement and body co-ordination. There is an intimate relationship between muscle and bone as both contribute to creating a movement in the body.

You will have learnt from the skeletal system that bones and joints provide the leverage in a movement, but it is in fact the muscle which provides the pull upon the bone to effect the movement. The key to learning the anatomical position and action of muscles is to first learn the individual position of the bones. It is then a logical step to learn the muscle attachments in relation to bone and what movements those muscles create. This chapter concentrates largely on the superficial muscles of the face and body as these are the muscles which cover the body and therefore are the ones upon which a therapist will be primarily working.

A competent therapist needs to be able to:

- have a good knowledge of the superficial muscles of the face and body in order to be able to carry out treatments safely and effectively

By the end of this chapter, you will be able to recall and understand the following in relation to your practical work:

- The structure of voluntary muscle tissue.
- Muscle contraction
- Muscle fatigue
- The effects of temperature and increased circulation on muscle contraction

- Muscle tone
- The anatomical position and action of the main superficial muscles of the face and body

The muscular system consists largely of skeletal muscle tissue which covers the bones on the outside, and connective tissue which attaches muscles to the bones of the skeleton. Muscles, along with connective tissue, help to give the body its contoured shape.

The muscular system has three main functions:

MOVEMENT

Consider the action of picking up a pen that has dropped onto the floor. This seemingly simple action of retrieving the pen involves the co-ordinated action of several muscles pulling on bones at joints to create movement. Muscles are also involved in the movement of body fluids such as blood, lymph and urine. Consider also the beating of the heart which is continuous throughout life.

MAINTAINING POSTURE

Some fibres in a muscle resist movement and create slight tension in order to enable us to stand upright. This is essential, as without body posture we would be unable to maintain normal body positions such as sitting down or standing up.

THE PRODUCTION OF HEAT

As muscles create movement in the body they generate heat as a by product, which helps to maintain our normal body temperature.

MUSCLE TISSUE

Muscle tissue makes up about 50% of your total body weight and is composed of:

20% protein
75% water
5% mineral salts, glycogen and fat

There are three types of muscle tissue in the body:

1 **Skeletal** or voluntary muscle tissue which is primarily attached to bone
2 **Cardiac** muscle tissue which is found in the walls of the heart
3 **Smooth** or involuntary muscle tissue which is found inside the digestive and urinary tracts, as well as in the walls of blood vessels

All three types of muscle tissue differ in their structure and functions and the degree of control the nervous system has upon them.

A therapist is primarily concerned with the structure of voluntary muscle tissue.

THE STRUCTURE OF VOLUNTARY MUSCLE TISSUE

Voluntary muscle tissue is made up of bands of elastic or contractile tissue bound together in bundles and enclosed by a connective tissue sheath, which protects the muscle and helps to give it a contoured shape. The ends of the sheath extend to form tendons, by which voluntary muscles are attached to bone.

—— K E Y N O T E – T E N D O N ——

A tendon is made up of white fibrous tissue, which attaches the muscle to the vascular membrane of a bone.

Voluntary muscles have many nuclei situated on their outer membrane. In microscopic structure, they are known to have a large number of striated fibres; this is because the contractile fibres that form them are connected in such a way that they appear to be striped. These contractile fibres or myofibrils in skeletal muscle run longitudinally and consist of two kinds of protein filaments:

- actin, which is the thinner filament
- myosin, which is the thicker filament

The two types of filaments are arranged in alternating bands, hence they appear *striped* or *striated*. These protein filaments are significant in the mechanism of muscle contraction.

Voluntary muscle works intimately with the nervous system and will therefore only contract if a stimulus is applied to it via a motor nerve. Each muscle fibre receives its own nerve impulse so that fine and varied motions are possible. Voluntary muscles also have their own small stored supply of glycogen which is used as fuel for energy. Voluntary muscle tissue differs from other types of muscle tissue in that the muscles tire easily, and need regular exercise.

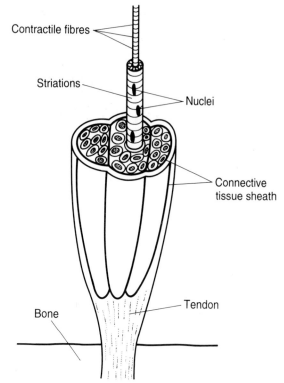

Figure 22
Structure of voluntary muscle tissue

MUSCLE CONTRACTION

Muscle tissue has several characteristics which help contribute to the functioning of a muscle:

- *Contractibility*: which is the capacity of the muscle to shorten and thicken
- *Extensibility*: which is the ability to stretch when the muscle fibres relax
- *Elasticity*: which is the ability to return to its original shape after contraction
- *Irritability*: which is the response to stimuli provided by nerve impulses

Muscles vary in the *speed* at which they contract. The muscle in your eyes will be moving very fast as you are reading this page, whilst the muscles in your limbs assisting you in turning the pages will be contracting at a moderate speed.

The speed of a muscle contraction is therefore modified to meet the demands of the action concerned and the degree of nervous stimulus it has received.

Skeletal or *voluntary* muscles are moved as a result of nervous stimulus which they receive from the brain via a motor nerve. Each nerve fibre ends in a motor point, which is the end portion of the nerve and is the part through which the stimulus to contract is given to the muscle fibre. A single motor nerve may transmit stimuli to one muscle fibre or as many as one hundred and fifty, depending on the effect of the action required.

THE CONTRACTION OF VOLUNTARY MUSCLE TISSUE

When a stimulus is applied to voluntary muscle fibres via a motor nerve a *mechanical* action is initiated:

- During contraction a sliding movement occurs within the contractile fibres of the muscle in which the actin protein filaments move inwards towards the myosin and the two filaments merge. This action causes the muscle fibres to shorten and thicken and then pull upon their attachments (bones and joints) to effect the movement required
- During relaxation, the muscle fibres elongate and return to their original shape

The force of muscle contraction depends upon the number of fibres in a muscle which contract simultaneously, and the more fibres involved, the stronger and more powerful the contraction will be.

THE ENERGY NEEDED FOR MUSCLE CONTRACTION

A certain amount of energy is needed to effect the *mechanical* action of the muscle fibres. This is obtained principally from carbohydrate foods such as glucose in the arterial blood supply. Glucose, which is not required immediately by the body, is converted into glycogen and is stored in the liver and the muscles. Muscle glycogen therefore provides the *fuel* for muscle contraction.

- During muscle contraction glycogen is broken down by a process called oxidation (where glucose combines with oxygen and releases energy). Oxygen is stored in the form of haemoglobin in the red blood cells and as myoglobin in the muscle cells
- During oxidation, a chemical compound called ATP (adenosine triphosphate) is formed. Molecules of ATP are contained within voluntary muscle tissue and their function is to temporarily store energy produced from food
- When the muscle is stimulated to contract, ATP is converted to another chemical compound, ADP (adenosine diphosphate), which releases the energy needed to be used during the phase of muscle contraction

- During the oxidation of glycogen, a substance called pyruvic acid is formed.
- If plenty of oxygen is available to the body, as in rest or undertaking moderate exercise, then the pyruvic acid is broken down into waste products, carbon dioxide and water which are excreted into the venous system. This is known as *aerobic* respiration
- If insufficient oxygen is available to the body as may be in the case of vigorous exercise then the pyruvic acid is converted into lactic acid. This is known as *anaerobic* respiration.

— K E Y   N O T E —

The waste product lactic acid, which diffuses into the bloodstream after vigorous exercise, causes the muscles to ache. This condition is known as muscle fatigue.

THE EFFECTS OF INCREASED CIRCULATION ON MUSCLE CONTRACTION

During exercise muscles require more oxygen to cope with the increased demands made on the body. During exercise the body is active in initiating certain circulatory and respiratory changes to the body to meet the increased oxygen requirements of the muscles.

CIRCULATORY CHANGES THAT OCCUR IN THE BODY DURING MUSCLE CONTRACTION

- During exercise, there is an increased return of venous blood to the heart, owing to the more extensive movements of the diaphragm and the general contractions of the muscles compressing the veins
- With the rate and output of each heart beat being increased, a greater volume of blood is circulated around the body, which will lead to an increase in the amount of oxygen in the blood
- More blood is distributed to the muscles and less to the intestine and skin to meet the needs of the exercising muscles. During exercise, a muscle may receive as much as 15 times its normal flow of blood.

RESPIRATORY CHANGES

- The presence of lactic acid in the blood stimulates the respiratory centre in the brain, increasing the rate and depth of breath, producing panting.

The rate and depth of breath remains above normal for a while after strenuous exercise has ceased; large amounts of oxygen are taken in to allow the cells of the muscles and the liver to dispose of the accumulated lactic acid by oxidising it and converting it to glucose or glycogen. Lactic acid is formed in the tissues in amounts far greater than can be immediately disposed of by available oxygen. The extra oxygen needed to remove the accumulated lactic acid is what is called the oxygen debt, which must be repaid after the exercise is over.

──────── K E Y   N O T E ────────

The conversion of lactic acid back into glucose is a relatively slow process and it may take several hours to repay the oxygen debt, depending on the extent of the exercise undertaken.

This situation can be minimised by massaging muscles before and after an exercise schedule, which will increase the blood supply to the muscles and prevent an excess of lactic acid forming in the muscles.

THE EFFECTS OF TEMPERATURE ON MUSCLE CONTRACTION

Exercising muscles produces heat, which is carried away from the muscle by the bloodstream and is distributed to the rest of the body. Exercise is, therefore, an effective way to increase body temperature. When muscle tissue is warm, the process of contraction will occur faster due to the acceleration of the chemical reactions and the increase in circulation. However, it is possible for heat cramps to occur in muscles which are exercised at high temperatures, as increased sweating causes loss of sodium in the body, leading to a reduction in the concentration of sodium ions in the blood supplying the muscle.

Cramp occurs when muscles become over-contracted and hence go into spasm; this is usually caused by an irritated nerve or an imbalance of mineral salts such as sodium in the body. Cramp most commonly affects the calf muscles or the soles of the feet. Cramp can be very painful as it is a sudden involuntary contraction of the muscle.

Treatment to relieve the pain of cramp includes stretching the affected muscle group and using soothing effluerage movements to help to relax the muscles.

Conversely, as muscle tissue is cooled, the chemical reactions and circulation slow, causing the contraction to be slower. This causes an involuntary increase in muscle tone known as shivering, that increases body temperature in response to cold.

MUSCLE TONE

Even in a relaxed muscle a few muscle fibres remain contracted to give the muscle a certain degree of firmness. At any given time, a small number of motor units in a muscle are stimulated to contract and cause tension in the muscle rather than full contraction and movement, whilst the others remain relaxed. The group of motor units functioning in this way change periodically so that muscle tone is maintained without fatigue. This state of partial contraction of a muscle is known as muscle tone and is important for maintaining body posture.

Muscles with less than the normal degree of tone are said to be *flaccid* and when the muscle tone is greater than normal the muscles become *spastic* and rigid.

─ KEY NOTE ─

Muscle tone will vary from client to client and will largely depend on the amount of exercise undertaken. Muscles with good tone have a better blood supply as their blood vessels will not be inhibited by fat.

❓ SELF-ASSESSMENT QUESTIONS (A)

1. Briefly describe the structure of voluntary muscle tissue

....

....

....

....

....

....

2. What is a tendon?

....

....

3. What causes a voluntary muscle to contract?

....

....

....

4. What is the body's principal carbohydrate needed for muscle contraction?

....

5. What is muscle fatigue?

....

....

....

....

....

6. What is muscle tone?

....

....

....

....

MUSCLES OF THE FACE AND NECK

◖◗ TASK I

Complete the following by identifying the name of the muscle.

TABLE 7: *Face and neck muscles*

Name	Location	Function	Key note
	lies obliquely across side of neck	flexes the neck, rotates head to one side	this muscle is used when nodding
	covers anterior of neck, extending from chest to chin	depresses mandible and lower lip	this muscle wrinkles the skin on the neck in old age

TABLE 7: *continued*

Name	Location	Function	Key note
	extends from lower lip to centre of chin	elevates lower lip and wrinkles chin	this muscle is used when expressing displeasure and when pouting
	between zygomatic bone and angle of mandible	elevates and closes jaw, assists in chewing	this muscle is used when gritting the teeth and therefore holds a lot of tension in the face
	forms main part of cheek between upper and lower jaw	compresses the cheeks, aids in chewing	this muscle is used in blowing
	surrounds the mouth	closes the lips, shapes the lips during speech	this muscle is used in puckering the lips to kiss and in whistling
	triangular shaped muscle, lies horizontally on cheek joins at corners of mouth	draws corners of mouth laterally	this muscle is used in grinning
	surrounds the eye	closes the eyes tightly	this muscle is used in blinking and winking

⦂◆ TASK 2

Label the superficial muscles of the face and neck on the following diagram.

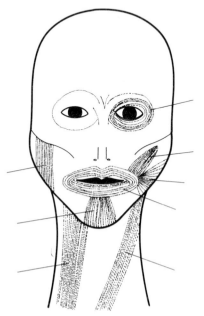

Figure 23
Muscles of the face and neck

❓ SELF-ASSESSMENT QUESTIONS (B)

1. Name the muscle that elevates the lower lip and wrinkles the chin

   ....

2. What is the action of the masseter muscle?

   ....

   ....

3. What is the action of the sternomastoid muscle?

   ....

   ....

4. Which muscle forms the main part of the cheek? What is its action?

....

....

....

....

5. What is the position and action of the obicularis oris muscle?

....

....

....

MUSCLES OF THE BACK

TASK 3

Complete the following table by identifying the name of the muscle.

TABLE 8: *Back muscles*

Name	Location	Function	Key note
	(broad sheet of tendon), extends across posterior of lower thorax	draws shoulder downwards and backwards, adducts and inwardly rotates arm	this muscle is used when climbing and rowing; when walking on crutches it helps to support weight of body on hands
	group of three bands of muscle, lie in a groove between spinal column and ribs and extend from sacrum to occipital bone of skull	extends spine, helps hold body upright	this muscle is very important to posture

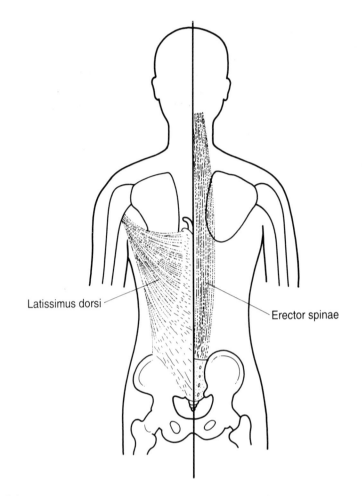

Figure 24
Muscles of the back

MUSCLES OF THE BUTTOCKS

✺ TASK 4

Complete the following table by identifying the name of the muscle.

TABLE 9: *Muscles of the buttocks*

Name	Location	Function	Key note
	posterior of hip, forming the buttocks	abducts thigh, laterally rotates thigh, extends hip joint	this muscle is used when running and jumping
	lateral aspect of buttocks between gluteus maximus and gluteus minimus	abducts thigh, rotates thigh medially	this muscle is used in walking and running and is used in balance and standing on one leg
	small muscle on lateral aspect of buttock, deep to gluteus medius	abducts thigh, laterally rotates thigh	this muscle is used in balance when standing on one leg

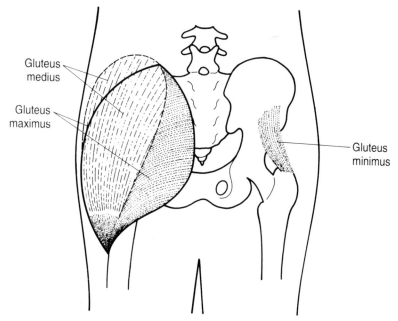

Gluteus medius

Gluteus maximus

Gluteus minimus

Figure 25
Muscles of the buttocks

MUSCLES OF THE CHEST

TASK 5

Complete the following table by identifying the name of the muscle.

TABLE 10: *Chest muscles*

Name	Location	Function	Key note
	triangular muscle, covers anterior surface of chest	adducts arm, inwardly rotates arm	this muscle is used in throwing
	small muscle, immediately deep to pectoralis major	draws shoulder downwards and forwards	this muscle is used when rolling the shoulders

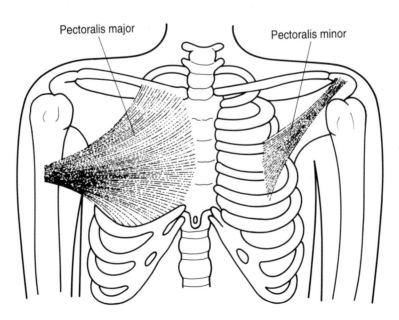

Figure 26
Muscles of the chest wall

MUSCLES OF THE ABDOMINAL WALL

TASK 6

Complete Table 11 by identifying the name of the muscle.

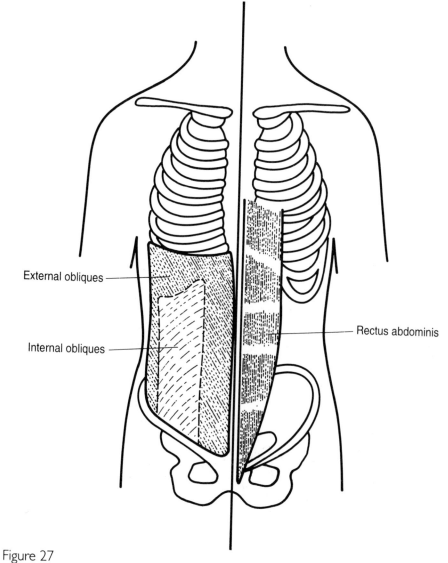

External obliques

Internal obliques

Rectus abdominis

Figure 27
Muscles of the abdominal wall

TABLE 11: *Muscles of the abdominal wall*

Name	Location	Function	Key note
	medially on anterior of abdomen, extends from pubis to bottom section of sternum	flexes trunk, flexes spine, compresses abdomen	this muscle is used when doing a sit-up and is an important postural muscle
	laterally from side of waist down anterior of abdomen	twists trunk to opposite side	this muscle runs in the direction in which you put your hands in your pockets
	laterally on anterior of abdomen, fibres run at right angles to external obliques	twists trunk to opposite side	the fibres of this muscle run at right angles to the external obliques

MUSCLES OF THE THORAX

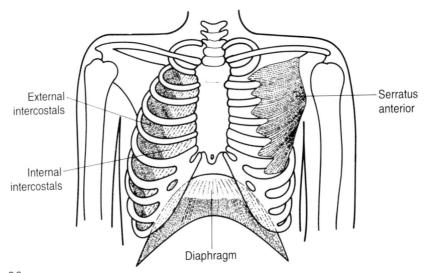

External intercostals

Serratus anterior

Internal intercostals

Diaphragm

Figure 28
Muscles of the thorax

◆ TASK 7

Complete the following table by identifying the name of the muscle.

TABLE 12: *Muscles of the thorax*

Name	Location	Function	Key note
	sides of ribcage below axilla	draws scapula forwards, rotates scapula upwards	this muscle is used when pushing and punching
	large dome-shaped muscle, separates thorax from abdomen	contracts and flattens during inspiration, returns to original dome-shape during expiration	the contraction of this muscle increases the volume of the thoracic cavity
	situated in between the ribs	elevates ribs during inspiration	the contraction of these muscles increases the depth of the thoracic cavity
	situated in between the ribs	depresses ribs during forced expiration	these muscles are used during forced expiration, i.e. when blowing a trumpet or coughing

MUSCLES OF THE SHOULDER

✏ TASK 8

Complete the following table by identifying the name of the muscle.

TABLE 13: *Shoulder muscles*

Name	Location	Function	Key note
	triangular muscle shaped like a trapezium, extends across posterior surface of neck and shoulders to mid-spine	elevates shoulder girdle	this muscle is used when shrugging the shoulders and is the first muscle to tighten when tense
	caps the top of the humerus and the shoulder joint	abduction of the arm	this muscle gives the rounded shape of the shoulder

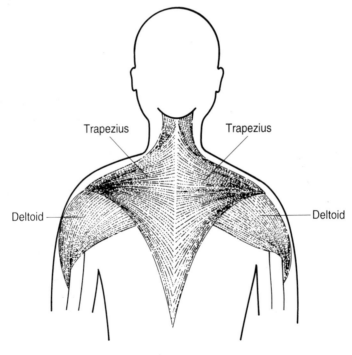

Figure 29
Muscles of the shoulder

MUSCLES OF THE UPPER ARM

◉ **TASK 9**

Complete the following table by identifying the name of the muscle.

TABLE 14: *Muscles of the upper arm*

Name	Location	Function	Key note
	anterior of upper arm	flexion and supination of forearm	this muscle has two heads and works in opposition to the triceps muscle
	distal half of anterior surface of humerus	flexes forearm	this muscle lies deep to the biceps muscle
	posterior of humerus	extension of forearm	this muscle works in opposition to the biceps muscle

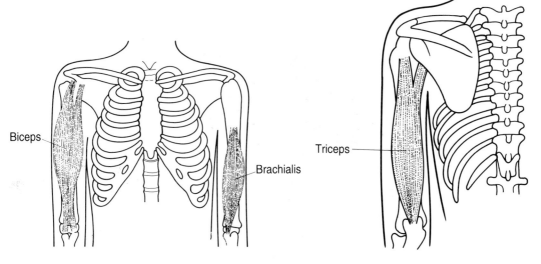

Figure 30
Muscles of the anterior of the upper arm

Figure 31
Muscles of the posterior of the upper arm

MUSCLES OF THE ANTERIOR OF THE THIGH

☞ TASK 10

Complete the following table by identifying the name of the muscle.

TABLE 15: *Muscles of the anterior of the thigh*

Name	Location	Function	Key note
	group of four muscles situated on the anterior of the thigh	extension of the knee joint	these muscles are used in walking, kicking and raising the body from a sitting or squatting position
	crosses anterior of thigh from ilium of pelvis to tibia	flexes thigh and medially rotates thigh	this muscle is used when playing football, e.g. dribbling a ball

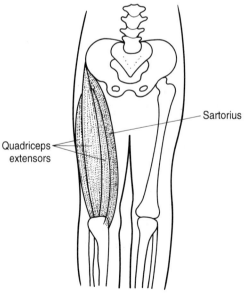

Figure 32
Muscles of the anterior of the thigh

MUSCLES OF THE MEDIAL AND LATERAL ASPECTS OF THE THIGH

✇ TASK 11

Complete the following table by identifying the names of the muscles.

TABLE 16: *Muscles of the medial and lateral aspects of thigh*

Name	Location	Function	Key note
	a group of four muscles situated on the medial side of the thigh	adduct and flex the thigh at the hip	these muscles are important in the maintenance of body posture: groin strains are a common problem associated with these muscles
	situated on the lateral side of the thigh	abducts the thigh and inwardly rotates thigh	this muscle is attached to a broad sheet of connective tissue which helps to strengthen the knee joint when walking and running

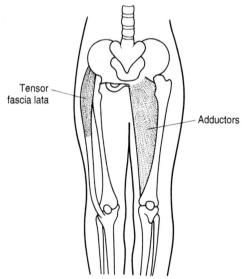

Figure 33

Muscles of the medial and lateral aspects of the thigh

MUSCLES OF THE POSTERIOR OF THE LOWER LEG AND THIGH

◄► TASK 12

Complete the following table by identifying the names of the muscles. You can refer to Figure 34 to help you with this task.

TABLE 17: *Muscles of the posterior of the lower leg and thigh*

Name	Location	Function	Key note
	a group of three muscles situated on the posterior of the thigh	flexes the knee and extends the hip	these muscles contract powerfully when raising the body from a stooped position, and when climbing stairs
	largest and most superficial muscle on calf	plantar flexes foot, flexes leg at knee	this muscle provides the push during fast walking and running
	deep to gastrocnemius, covers whole of calf area	assists in plantar flexing of the foot	this muscle is flatter and thicker than gastrocnemius and is important as a postural muscle

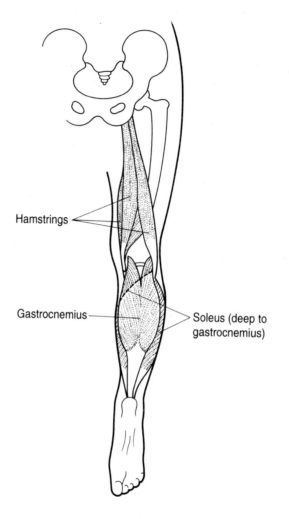

Hamstrings

Gastrocnemius

Soleus (deep to gastrocnemius)

Figure 34
Muscles of the posterior of the lower leg and thigh

MUSCLES OF THE ANTERIOR OF THE LOWER LEG

⊷ TASK 13

Complete the following table by identifying the name of the muscle.

TABLE 18: *Muscles of the anterior of the lower leg*

Name	Location	Function	Key note
	anterior of lower leg, runs alongside shaft of tibia	dorsiflexes and inverts the foot	this muscle is used when taking the foot off the clutch pedal when driving

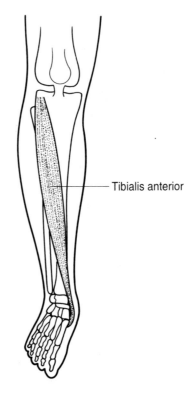

— Tibialis anterior

Figure 35
Muscles of the anterior of the lower leg

SUPERFICIAL MUSCLES – ANTERIOR OF BODY

☛ TASK 14

Label the superficial muscles of the anterior of the body on the following diagram.

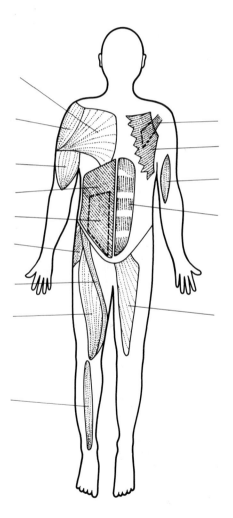

Figure 36
Superficial muscles – anterior of body

SUPERFICIAL MUSCLES – POSTERIOR OF BODY

TASK 15

Label the superficial muscles of the posterior of the body on the following diagram.

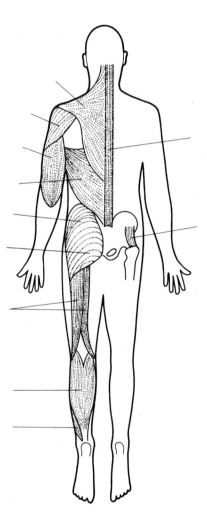

Figure 37
Superficial muscles – posterior of body

? SELF ASSESSMENT QUESTIONS (C) – SUPERFICIAL MUSCLES OF THE BODY

1. What is the action of the gluteus maximus muscle?

   ....

   ....

   ....

2. Name the muscle which extends the spine

   ....

   ....

3. What is the position of the pectoralis major muscle?

   ....

   ....

4. What is the action of the rectus abdominus muscle?

   ....

   ....

5. Name the muscles involved in respiration

   ....

   ....

   ....

6. What is the action of the trapezius muscle?

   ....

   ....

7. What is the action of the biceps and the triceps muscles of the upper arm?

   ....

   ....

8. Name the group of muscles which extend the knee joint

   ....

   ....

9. What is the action of the hamstring muscles?

   ....

   ....

10. What is the position and action of the gastrocnemius muscle?

   ....

   ....

   ....

Chapter 7

THE CIRCULATORY SYSTEM

The Circulatory system is the body's transport system and comprises blood, blood vessels and the heart. Blood provides the fluid environment for our body's cells and it is transported in specialised tubes called blood vessels. The heart acts like a pump which keeps the blood circulating around the body in a constant circuit.

A competent therapist needs to be able to understand:

- the principles of circulation to understand the physiological effects of treatments and to be able to carry out treatments safely and effectively

By the end of this chapter, you will be able to relate the following knowledge to your practical work carried out in the salon:

- The structural and functional differences between the different blood vessels
- The major blood vessels of the heart
- The pulmonary and systemic blood circulation
- Blood pressure and the pulse rate
- Circulatory disorders such as high or low blood pressure and varicose veins

Blood is the medium in which all materials are carried to and from the cells of the body. Blood, therefore, provides the fluid environment for cells.

FUNCTIONS OF BLOOD

There are four main functions of blood.

1. TRANSPORT

- oxygen is carried from the lungs to the cells of the body in red blood cells
- carbon dioxide is carried from the body's cells to the lungs
- nutrients such as glucose, amino acids, vitamins and minerals are carried from the small intestine to the cells of the body
- cellular wastes such as water, carbon dioxide, lactic acid and urea are carried in the blood to be excreted
- hormones, which are internal secretions that help to control important body processes, are transported by the blood to target organs

KEY NOTE

Red blood cells are called erythrocytes and they contain the red protein pigment haemoglobin that combines with oxygen to form oxyhaemoglobin. The pigment haemoglobin assists the function of the erythrocyte in transporting oxygen from the lungs to the body's cells and carrying carbon dioxide away.

2. DEFENCE

- white blood cells are collectively called leucocytes and they play a major role in combating disease and fighting infection

KEY NOTE

White blood cells are known as phagocytes as they have the ability to engulf and ingest micro-organisms which invade the body and cause disease. Specialised white blood cells called lymphocytes produce antibodies to protect the body against infection.

3. REGULATION

- blood helps to regulate heat in the body by absorbing large quantities of heat produced by the liver and the muscles; this is then transported around the body to help to maintain a constant internal temperature
- blood also helps to regulate the body's pH balance

4. CLOTTING

- if the skin becomes damaged, specialised blood cells called platelets clot to prevent the body from losing too much blood and to prevent the entry of bacteria

BLOOD VESSELS

Blood flows round the body by the pumping action of the heart and is carried in vessels known as arteries, veins and capillaries.

KEY FACTORS ABOUT BLOOD VESSELS

ARTERIES

- Arteries carry blood away from the heart
- Blood is carried under high pressure
- Arteries have thick muscular and elastic walls to withstand pressure
- Arteries have no valves, except at the base of the pulmonary artery, where they leave the heart
- Arteries carry oxygenated blood, except the pulmonary artery to the lungs
- Arteries are generally deep-seated, except where they cross over a pulse spot
- Arteries give rise to small blood vessels called arterioles, which deliver blood to the capillaries

VEINS

- Veins carry blood towards the heart
- Blood is carried under low pressure
- Veins have less thick, muscular walls
- Veins have valves at intervals to prevent the back flow of blood
- Veins carry deoxygenated blood, except the pulmonary veins from the lung
- Veins are generally superficially, not deep seated
- Veins form finer blood vessels called venules which continue from capillaries

CAPILLARIES

- Capillaries are the smallest vessels
- Capillaries unite arterioles and venules, forming a network in the tissues
- The wall of a capillary vessel is only a single layer of cells thick, it is therefore sufficiently thin to allow the process of diffusion of dissolved substances to and from the tissues to occur
- Capillaries have no valves

- Blood is carried under low pressure, but higher than in veins
- Capillaries are responsible for supplying the cells and tissues with nutrients

---

KEY NOTE
– CAPILLARY EXCHANGE

The key function of a capillary is to permit the exchange of nutrients and waste between the blood and tissue cells. Substances such as oxygen, vitamins, minerals and amino acids pass through to the tissue fluid to nourish the nearby cells, and substances such as carbon dioxide and waste are passed out of the cell. This exchange of nutrients can only occur through the semi-permeable membrane of a capillary, as the walls of arteries and veins are too thick.

Oxygenated blood flowing through the arteries appears bright red in colour due to the oxygen pigment haemoglobin; as it moves through capillaries it off loads some of its oxygen and picks up carbon dioxide. This explains why blood flow in veins appears darker.

STRUCTURAL AND FUNCTIONAL DIFFERENCES BETWEEN ARTERIES, VEINS AND CAPILLARIES

✦ TASK I

Complete the following table to show the structural and functional differences between arteries, veins and capillaries.

TABLE 19: *Structural and functional differences between arteries, veins and capillaries*

	Walls	**Valves**	**Direction of blood flow**	**Blood pressure**	**Function**
Artery	thick, muscular	none	away from heart	high	
Vein	thick, muscular	numerous	towards heart	low	
Capillary	thin single cell layer	none	arteriole–venule	high–low	

THE HEART

The heart is a hollow organ made up of cardiac muscle tissue which lies in the thorax above the diaphragm and between the lungs. It is composed of three layers of tissue:

PERICARDIUM: THE OUTER LAYER

This is a double-layered bag enclosing a cavity filled with pericardial fluid, which reduces friction as the heart moves during its beating.

MYOCARDIUM: THE MIDDLE LAYER

This is a strong layer of cardiac muscle which makes up the bulk of the heart.

ENDOCARDIUM: THE INNER LAYER

This lines the heart's cavities and is continuous with the lining of the blood vessels.

The heart is divided into a right and left side by a partition called a septum and each side is further divided into a thin-walled atrium above and a thick-walled ventricle below. Between the right atrium and the right ventricle is the tricuspid valve and between the left atrium and the left ventricle is the bicuspid or mitral valve. These valves help to maintain the direction of blood flow through the heart. The heart muscle is supplied by the right and left coronary blood vessels.

FUNCTION

The function of the heart is to maintain a constant circulation of blood throughout the body. The heart acts as a pump and its action consists of a series of events known as the cardiac cycle.

Blood is transported as part of a double circuit and consists of two separate systems which are joined only at the heart.

THE PULMONARY CIRCULATION

This consists of the circulation of deoxygenated blood from the right ventricle of the heart to the lungs, where it becomes oxygenated and is then returned to the left atrium by the pulmonary veins, to be passed to the aorta for the general or systemic circulation.

The pulmonary circulation is essentially the circulatory system between the heart and the lungs where a high concentration of blood oxygen is restored and the concentration of carbon dioxide in the blood is lowered.

THE GENERAL OR SYSTEMIC CIRCULATION

The systemic circuit is the largest circulatory system and carries oxygenated blood from the left ventricle of the heart through to the aorta. Oxygenated blood is then passed around the body through the various branches of the aorta and deoxygenated blood is returned to the right atrium via the superior and inferior vena cava.

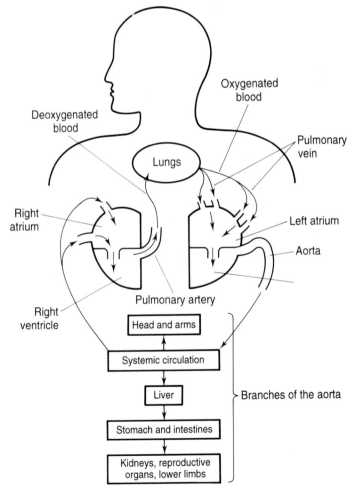

Figure 38
Pulmonary and systemic circulation

The aorta is divided into three main branches which subdivide into branches which supply the whole of the body:

- The ascending arch has branches which supply the head, neck and the top of the arms
- The descending thoracic has branches which supply organs of the thorax
- The descending abdominal has branches which supply the legs and organs of the digestive, renal and reproductive systems

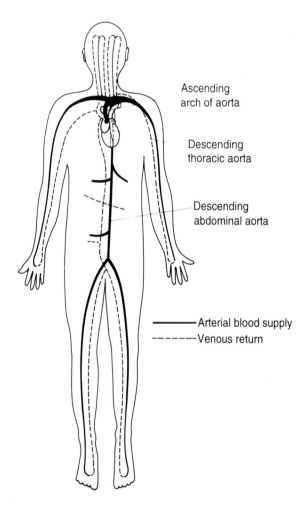

Ascending
arch of aorta

Descending
thoracic aorta

Descending
abdominal aorta

————Arterial blood supply
- - - - - -Venous return

Figure 39
Branches of the aorta

KEY NOTE
– BLOOD SHUNTING

Along certain circulatory pathways such as in the intestines there are strategic points where small arteries have direct connection with veins. When these connections are open they act as shunts which allow blood in the artery to have direct access to a vein. These interconnections allow for sudden and major diversions of blood volume according to the physical needs of the body. In relation to circulation, this means that treatment should not be given after a heavy meal due to the increased circulation to the intestines, resulting in a diminished supply to other areas of the body.

BLOOD FLOW TO THE HEAD AND NECK

TASK 2

Label the blood vessels supplying the head and neck on the following diagrams.

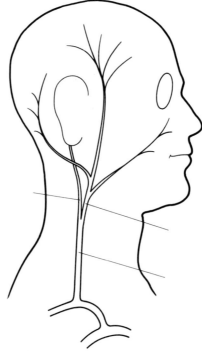

Figure 40
Arterial blood supply to the head and neck

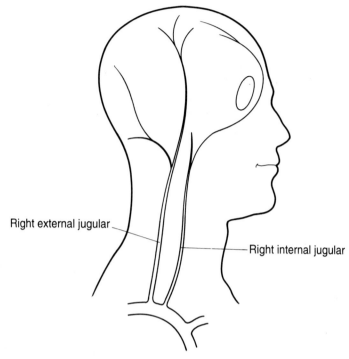

Right external jugular

Right internal jugular

Figure 41
Venous return from the head and neck

MAIN BLOOD VESSELS

TASK 3

Complete the following by identifying the name of the main blood vessel involved in blood circulation.

TABLE 20: *Main blood vessels*

Main blood vessel	Description
	receives deoxygenated blood from upper part of body and empties into right atrium of heart
	receives deoxygenated blood from lower part of body and empties into right atrium of heart
	carries deoxygenated blood from right ventricle of heart to lungs
	carries oxygenated blood from lungs back to the heart

TABLE 20: *continued*

Main blood vessel	Description
	main artery of the body, carries oxygenated blood around the body through its branches
	passes through temporal bone of skull to supply oxygenated blood to brain, eyes, forehead and nose
	divides into branches, to supply oxygenated blood to muscles and skin of face, sides and back of head
	forms major venous drainage to head and neck, collects deoxygenated blood from brain
	collects deoxygenated blood from superficial regions of face, scalp and neck

Blood pressure is the force with which blood is pumped through the arteries in order to push it through to the smaller vessels of circulation. The pressure in the arteries varies during each heartbeat. The maximum pressure of the heartbeat is known as the systolic pressure and can be measured when the heart muscle contracts and pushes blood out into the body through the arteries. The minimum pressure is when the heart muscle relaxes and blood flows into the heart from the veins. This is known as the diastolic pressure. Blood pressure may be measured with the use of a sphygmomanometer.

FACTORS AFFECTING BLOOD PRESSURE

Because blood pressure is the result of the pumping of the heart in the arteries, anything that makes the heart beat faster will raise the blood pressure. Factors affecting the blood pressure include:

- excitement
- anger
- stress
- fright
- pain
- exercise
- smoking and drugs

A normal blood pressure reading is between 100 and 140 mmHg systolic and between 60 and 90 mmHg diastolic. Blood pressure is measured in millimetres of mercury and is expressed as 120/80 mmHg.

THE PULSE

The pulse is a pressure wave that can be felt in the arteries which corresponds to the beating of the heart. The pumping action of the left ventricle of the heart is so strong that it can be felt as a pulse in arteries a considerable distance from the heart. The pulse can be felt at any point where an artery lies near the surface. The radial pulse can be found by placing two or three fingers over the radial artery below the thumb. Other sites where the pulse may be felt include the carotid artery at the side of the neck and over the brachial artery at the elbow.

The average pulse in an adult is between 60 and 80 beats per minute. Factors affecting the pulse rate include:

- exercise
- heat
- strong emotions such as grief, fear, anger or excitement

DISORDERS OF THE CIRCULATORY SYSTEM

High blood pressure is when the resting blood pressure is above normal. The World Health Organisation defines high blood pressure as consistently exceeding 160 mmHg systolic and 95 mmHg diastolic. High blood pressure is a common complaint and if serious may result in a stroke or a heart attack, due to the fact that the heart is made to work harder to force blood through the system. Causes of high blood pressure include:

- smoking
- obesity
- lack of regular exercise
- eating too much salt
- excessive alcohol consumption
- too much stress

High blood pressure can be controlled by:

- anti-hypertensive drugs which help to regulate and lower blood pressure
- decreasing salt and fat intake to prevent hardening of the arteries
- keeping weight down
- giving up smoking and cutting down on alcohol consumption
- relaxation and leading a less stressful life

Low blood pressure is when the blood pressure is below normal and is defined by the World Health Organisation as a systolic blood pressure of 99 mmHg or less and a diastolic of less than 59 mmHg.

Low blood pressure may be normal for some people in good health, during rest and after fatigue. The danger with low blood pressure is an insufficient supply of blood reaching the vital centres of the brain. Treatment may be by medication, if necessary.

KEY NOTE

High and low blood pressure are normally contra-indicated to treatments but with GP referral and an adaptation of routine, treatment may be possible. Correct positioning of the couch is essential to maximise comfort of the client with blood pressure problems and care needs to be taken to ensure that they are not lying down too long or get up too fast.

VARICOSE VEINS

Veins are known as varicose when the valves within them lose their strength. As a result of this, blood flow may become reversed or static. Valves are concerned with preventing the back flow of blood, but when their function is impaired they are unable to prevent the blood from flowing downwards, hence the walls of the affected veins swell and bulge out and become visible through the skin. Varicose veins may be due to several factors:

• hereditary tendencies
• ageing
• obesity, as excess weight puts pressure on the walls of the veins
• pregnancy
• sitting or standing motionless for long periods of time, causing pressure to build up in the vein

---

KEY NOTE

Varicose veins can be extremely painful and great care needs to be taken with a client with varicose veins. Treatment is therefore contra-indicated in the area affecting the veins.

? SELF-ASSESSMENT QUESTIONS

I. List the functions of blood

....

....

....

....

....

....

2. What is the function of an artery in blood transport?

....

....

3. What is the function of a vein in blood transport?

....

....

4. Why is a capillary one cell layer thick?

....

....

5. What is meant by
   a) the pulmonary circulation?
   b) the systemic circulation?
   Name the blood vessels involved and state whether they carry oxygenated or deoxygenated
   blood

....

....

....

....

....

6. What is meant by the term blood pressure?

....

....

....

7. List four factors affecting blood pressure

....

....

....

....

8. What is the pulse?

....

....

....

....

9. What is meant by high blood pressure and how does this affect salon treatment?

....

....

....

....

10. What are varicose veins and how do they affect salon treatment?

....

....

....

Chapter 8

THE LYMPHATIC SYSTEM

The lymphatic system is a one-way drainage system for the tissues, as it helps to provide a circulatory pathway for tissue fluid to be transported, as lymph, from the tissue spaces of the body into the venous system, where it becomes part of the blood circulation.

A competent therapist needs to be able to:

- understand the connection between blood and lymph in order to carry out treatments effectively

By the end of this chapter you will be able to relate the following knowledge to your practical work carried out in the salon:

- What lymph is and how it is formed
- The connection between blood and lymph
- The circulatory pathway of lymph
- The names and positions of the main lymph nodes of the head, neck and the body
- The drainage of lymph from the head, neck and the body
- The functions of the lymphatic system

WHAT IS LYMPH?

Lymph is a transparent, colourless, watery fluid which is derived from tissue fluid and is contained within lymph vessels. It resembles blood plasma, except that it has a lower concentration of plasma proteins. This is because some large protein molecules are unable to filter through the cells forming the capillary walls so they remain in blood plasma. Lymph contains only one type of cell; these are called lymphocytes.

HOW IS LYMPH FORMED?

As blood is distributed to the tissues some of the plasma escapes from the capillaries and flows around the tissue cells, delivering nutrients such as oxygen and water to the cell and picking up cellular waste such as urea and carbon dioxide. Once the plasma is outside the capillary and is bathing the tissue cells it becomes tissue fluid.

Some of the tissue fluid passes back into the capillary walls to return to the blood stream via the veins and some is collected up by a lymph vessel where it becomes lymph. Lymph is then taken through its circulatory pathway and is ultimately returned to the bloodstream.

THE CONNECTION BETWEEN BLOOD AND LYMPH

The lymphatic system is therefore often referred to as a secondary circulatory system as it consists of a network of vessels that assist the blood in returning fluid from the tissues back to the heart.

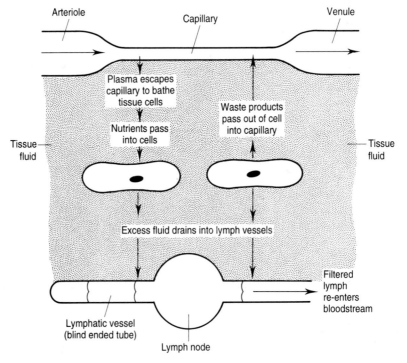

Figure 42
Connection between blood and lymph

The structures of the lymphatic system are as follows:

- lymph capillaries
- lymph vessels
- lymph nodes
- lymph collecting ducts

LYMPH CAPILLARIES

Lymph capillaries commence in the tissue spaces of the body as minute blind-end tubes, as lymph is a one-way circulatory pathway. The walls of the lymph capillaries are like those of the blood capillaries in that they are a single cell layer thick to make it possible for tissue fluid to enter them; however, they are permeable to substances of larger molecular size than those of the blood capillaries.

The lymph capillaries mirror the blood capillaries and form a network in the tissues draining away excess fluid and waste products from the tissue spaces of the body. Once the tissue fluid enters a lymph capillary it becomes lymph and is gathered up into larger lymph vessels.

——— KEY NOTE – OEDEMA ———

The term oedema refers to an excess of fluid within the tissue spaces that causes the tissues to become waterlogged.

LYMPH VESSELS

Lymph vessels are similar to veins in that they have thin collapsible walls and their role is to transport lymph through its circulatory pathway. They have a considerable number of valves which help to keep the lymph flowing in the right direction and prevent back flow. Superficial lymph vessels tend to follow the course of veins by draining the skin, whereas the deeper lymph vessels tend to follow the course of arteries and drain the internal structures of the body.

The lymph vessels carry the lymph towards the heart under steady pressure and about two to four litres of lymph pass into the venous system every day. Once lymph has passed through the lymph vessels it drains into at least one lymphatic node before returning to the blood circulatory system.

KEY NOTE

As the lymphatic system lacks a pump, lymphatic vessels have to make use of contracting muscles that assist the movement of lymph. Therefore, lymphatic flow is at its greatest during exercise due to the increased contraction of muscle.

LYMPH NODES

A lymph node is an oval or bean shaped structure, covered by a capsule of connective tissue. It is made up of lymphatic tissue and is divided into two regions: an outer cortex and an inner medulla.

There are more than one hundred lymph nodes, placed strategically along the course of lymph vessels. They vary in size between one millimetre and twenty five millimetres in length and are massed in groups; some are superficial and lie just under the skin, whereas others are deeply seated and are found near arteries and veins. Each lymph node receives lymph from several afferent lymph vessels and blood from small arterioles and capillaries. Valves of the afferent lymph vessels open towards the node, therefore lymph in these vessels can only move towards the node. Lymph flows slowly through the node, moving from the cortex to the medulla, and leaves through an efferent vessel which opens away from the node. The function of a lymph node is to act as a filter of lymph to remove or trap any micro-organisms, cell debris, or harmful substances which may cause infection, so that, when lymph enters the blood, it has been cleared of any foreign matter. When lymph enters a node, it comes into contact with two specialised types of leucocytes:

- monocytes, which are phagocytic in action; they engulf and destroy dead cells, bacteria and foreign material in the lymph
- lymphocytes, which are reproduced within the lymph nodes and can neutralise invading bacteria and produce chemicals and antibodies to help fight disease

---

KEY NOTE

If an area of the body becomes inflamed or otherwise diseased, the nearby lymph nodes will swell up and become tender, indicating that they are actively fighting the infection.

Once filtered, the lymph leaves the node by one or two efferent vessels, which open away from the node. Lymph nodes occur in chains, so that the efferent vessel of one node becomes the afferent vessel of the next node, in the pathway of lymph flow. Lymph drains through at least one lymph node before it passes into two main collecting ducts before it is returned to the blood.

☞ TASK 1

Complete the following table by identifying the name of the lymph nodes from the information given.

TABLE 21: *Major lymph nodes*

Name of node	Lymph drainage
	drains lymph from back of scalp and upper part of neck
	drains lymph from chin, lips, nose, cheek and tongue
	drains lymph from lower part of ear and cheek region
	drains lymph from larynx, oesophagus, posterior of scalp and neck, superficial part of chest and arm
	drains lymph from upper limbs and the chest wall
	drains lymph from the lower limbs and the abdominal walls
	drains lymph from abdominal organs
	drains lymph from the lower limbs through deep and superficial nodes

MAJOR LYMPH NODES OF HEAD AND NECK

TASK 2

Label the lymph nodes of the head and neck.

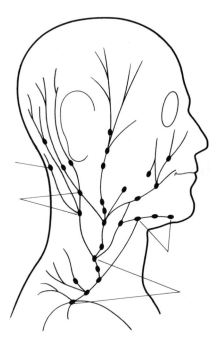

Figure 43
Major lymph nodes of the head and neck

MAJOR LYMPH NODES OF THE BODY

TASK 3

Label the major lymph nodes of the body.

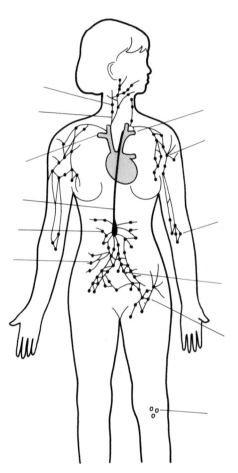

Figure 44
Major lymph nodes of the body

LYMPHATIC DUCTS

From each chain of lymph nodes, the efferent lymph vessels combine to form lymph trunks which empty into two main ducts: the thoracic and the right lymphatic ducts. These ducts collect lymph from the whole body and return it to the blood via the subclavian veins.

The thoracic duct: is the main collecting duct of the lymphatic system. It is the largest lymph vessel in the body and extends from the second lumbar vertebra up through the thorax to the root of the neck. The thoracic duct collects lymph from the left side of the head and neck, the left arm, the lower limbs and abdomen and drains into the left subclavian vein to return it to the bloodstream.

The right Lymphatic duct: is very short in length. It lies in the root of the neck and collects lymph from the right side of the head and neck and the right arm and drains into the right subclavian vein to be returned to the bloodstream.

SUMMARY OF THE CIRCULATORY PATHWAY OF LYMPH

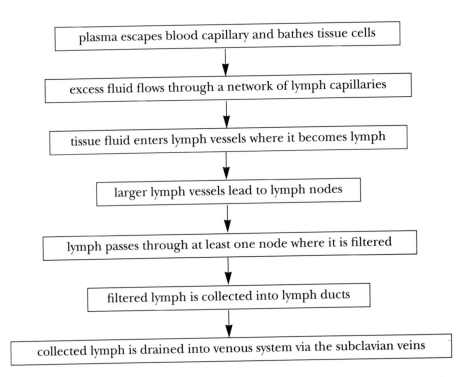

FUNCTIONS OF THE LYMPHATIC SYSTEM

- The lymphatic system is important for the distribution of fluid and nutrients in the body, because it drains excess fluid from the tissues and returns to the blood protein molecules, which are unable to pass back through the blood capillary walls because of their size
- The lymph nodes help to fight infection by filtering lymph and destroying invading micro-organisms. Lymphocytes are reproduced in the lymph nodes and following infection they generate antibodies to protect the body against subsequent infection. Therefore the lymphatic system plays an important part in the body's immune system
- The lymphatic System also plays an important part in absorbing the products of fat digestion from the villi of the small intestine. While the products of carbohydrate and protein digestion pass directly into the bloodstream, fats pass directly into the intestinal lymph vessels, known as *lacteals*

? SELF-ASSESSMENT QUESTIONS

1. What is lymph?

....

....

....

....

2. What is the connection between blood and lymph?

....

....

....

....

....

....

3. Complete the following in relation to the circulatory pathway of lymph
    i)   lymph capillaries
    ii) lymph...
    iii) lymph...
    iv) lymph ducts
    v)   filtered lymph returns to bloodstream via...................................   ....

4. Give a *brief* description of the structure of a lymph vessel

    ....

    ....

    ....

    ....

5. What is the function of a lymph node?

    ....

    ....

    ....

    ....

6. What are the functions of the lymphatic system?

    ....

    ....

    ....

    ....

    ....

    ....

    ....

    ....

    ....

Chapter 9

THE RESPIRATORY SYSTEM

The mechanism of respiration is the process by which the living cells of our body receive a constant supply of oxygen and remove carbon dioxide, and involves structures of the respiratory system, which are enclosed within the thoracic cavity.

A competent therapist needs to have knowledge of the mechanism of respiration in order to understand the importance of correct breathing.

By the end of this chapter you will be able to relate the following knowledge to your practical work carried out in the salon:

- the functional significance of the main structures of the respiratory system
- the process of the interchange of gases in the lungs
- the mechanism of breathing
- the importance of correct breathing

The respiratory system consists of the following structures which provide the passageway for air in and out of the body:

- The nose
- The naso-pharynx
- The pharynx
- The larynx
- The trachea
- The bronchi
- The lungs

THE NOSE

The nose is divided into the right and left cavities and is lined with tiny hairs called cilia, which begin to filter the incoming air, and mucous membrane which secretes a sticky fluid called mucus which helps to prevent dust and bacteria from entering the lungs. The nose moistens, warms and filters the air and is an organ which senses smell.

THE NASO-PHARYNX

The naso-pharynx is the upper part of the nasal cavity behind the nose, and is lined with mucous membrane. The naso-pharynx continues to filter, warm and moisten the incoming air.

THE PHARYNX

The pharynx or throat is a large cavity which lies behind the mouth and between the nasal cavity and the larynx. The pharynx serves as an air and food passage but cannot be used for both purposes at the same time, otherwise choking would result. The air is also warmed and moistened further as it passes through the pharynx.

THE LARYNX

The larynx is a short passage connecting the pharynx to the trachea and contains the vocal cords. The larynx has a rigid wall and is composed mainly of muscle and cartilage, which help to prevent collapse and obstruction of the airway. The larynx provides a passageway for air between the pharynx and the trachea.

THE TRACHEA

The trachea or windpipe is made up mainly of cartilage, which helps to keep the trachea permanently open. The trachea passes down into the thorax and connects the larynx with the bronchi, which pass into the lungs.

THE BRONCHI

The bronchi are two short tubes, similar in structure to the trachea, which lead to and carry air into each lung. They are lined with mucous membrane and ciliated cells and, like the trachea, contain cartilage to hold them open. The mucus traps solid particles and cilia move it upwards, preventing dirt from entering the delicate lung tissue. The bronchi subdivide into bronchioles in the lungs. These subdivide yet again and finally end in minute air-filled sacs called alveoli.

THE LUNGS

The lungs are cone-shaped spongy organs, situated in the thoracic cavity on either side of the heart. Internally the lungs consist of tiny air sacs called alveoli, which are arranged in lobules and resemble bunches of grapes. The function of the lungs is to facilitate the exchange of the gases oxygen and carbon dioxide, and in order to carry this out efficiently the lungs have several important features:

- a very large surface area provided by the alveoli
- thin permeable membrane surrounding the walls of the alveoli
- a thin film of water lining the alveoli, which is essential for dissolving oxygen from the alveoli air
- thin walled blood capillaries forming a network around the alveoli, which absorb oxygen from the air breathed into the lungs and release carbon dioxide into the air breathed out of the alveoli

The structures enclosed within the lungs are bound together by elastic and connective tissue. On the outside, the lungs have a serous covering or membrane called a pleura, which prevents friction between the lungs and the chest wall.

STRUCTURES OF THE RESPIRATORY SYSTEM

❖ TASK 1

Label the structures of the respiratory system on the following diagram.

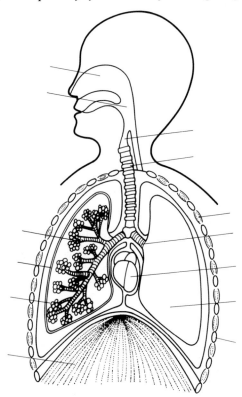

Figure 45
Structures of the respiratory system

THE INTERCHANGE OF GASES IN THE LUNGS

This process involves the absorption of oxygen from the air in exchange for carbon dioxide, which is released by the body as a waste product of cell metabolism:

- oxygen is taken in through the nose and mouth and flows along the trachea and bronchial tubes to the alveoli of the lungs, where it diffuses through the thin film of moisture lining the alveoli

- the inspired air, which is now rich with oxygen, comes into contact with the blood in the capillary network surrounding the alveoli
- the oxygen then diffuses across a permeable membrane wall surrounding the alveoli, to be taken up by red blood cells and carried to the heart
- carbon dioxide, collected by the respiring cells around the body, passes in the opposite direction by diffusing from the capillary walls into the alveoli, to be passed through the bronchi and trachea and exhaled through the nose and mouth

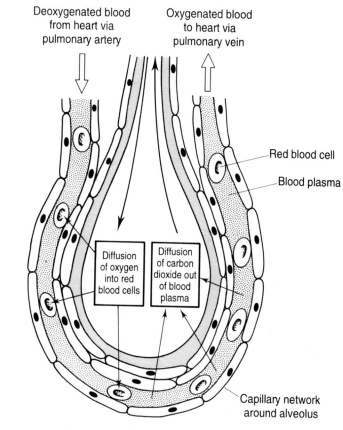

Figure 46
Interchange of gases

THE MECHANISM OF RESPIRATION

Air is moved in and out of the lungs by the combined action of the diaphragm and the intercostal muscles.

During *inspiration* or breathing the muscle fibre of the dome-shaped diaphragm contracts and becomes flatter by pushing down against the contents of the abdominal cavity. This increases the volume of the thoracic cavity and causes the lungs to fill with air. At the same time, the external intercostal muscles contract to increase the depth of the thoracic cavity, by pulling the ribs upwards and outwards.

During *expiration* air is breathed out due to the relaxation of the diaphragm and the external intercostal muscles, and the elastic recoil of the lungs. The diaphragm returns to its original shape, which causes the thoracic cavity to return to its original shape.

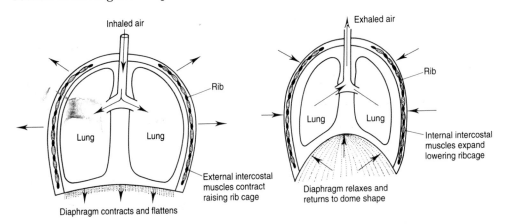

Figure 47
The mechanism of respiration

———————— KEY NOTE ————————

Breathing is a relatively passive process; however, when more air needs to be exhaled, such as when playing a wind instrument, the internal intercostal muscles contract to force more air out of the lungs. The internal intercostals may also be assisted by the muscles of the abdominal wall which will help to squeeze more air out of the lungs.

Exercise increases the rate and depth of breathing due to the muscle cells requiring more oxygen, and the breathing rate can more than double during vigorous exercise. Correct breathing is very important as it ensures that all the body's cells receive an adequate amount of oxygen and dispose of enough carbon dioxide to enable them to function efficiently.

? SELF-ASSESSMENT QUESTIONS

1. Name the main structures involved in respiration

....

....

....

....

....

....

....

....

2. Give a brief description of the process of the interchange of gases

....

....

....

....

....

....

....

....

3. Briefly describe the mechanism of respiration that causes air to be drawn in and out of the lungs

....

....

....

....

....

....

4. What is the importance of correct breathing?

....

....

....

....

....

....

....

Chapter 10

THE OLFACTORY SYSTEM

Olfaction is a special sense, which is capable of detecting different smells and evoking emotional responses due to its close link with the endocrine system. The process of olfaction is assisted by the nervous system, as smells received by the nose are transmitted by nerve impulses to be perceived by the brain.

A competent therapist needs to be able to have knowledge of the olfactory system in order to understand the process of olfaction

By the end of this chapter, you will be able to relate the following knowledge to your practical work carried out in the salon:

- The internal structures of the nasal cavity, along with the functional significance
- The theory of olfaction

The special features of the olfactory system are as follows:

- **the nose**: is the organ of olfaction or smell
- **mucous membrane**: lines the nose, moistens the air passing over it and helps to dissolve the odorous gas passing through the nasal cavity. The mucous membrane has a very rich blood supply, and warmth from the blood flowing through the tiny capillaries in the nose raises the temperature of the air as it passes through the nose
- **cilia**: are the tiny hairs inside the nose. They are highly sensitive and are extensions of nerve fibres connecting with the olfactory cells. The tips of the cilia are covered with mucous and they are able to detect tiny chemical odorous particles which enter the nose

- **olfactory cells**: which lie embedded in the mucous in the upper part of the nasal cavity. These nerve cells are sensory and are specially adapted for sensing smell. Each olfactory cell has a long nerve fibre called an axon, leading out of the main body of the cell, which picks up information received and passes it on to the brain
- **olfactory bulb**: the area of the brain, situated in the cerebral cortex, which perceives smell

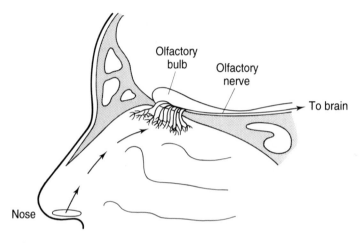

Figure 48
Internal structure of the nose

THE THEORY OF OLFACTION

- particles of solid or liquid (essential oil) evaporate on contact with the air
- mucous membrane, lining the nose, dissolves odorous particles by warming them and mixing them with water vapour as they pass through the nose
- special cilia pass on to olfactory cells whatever information they have picked up about the evaporated gas passing through the nose

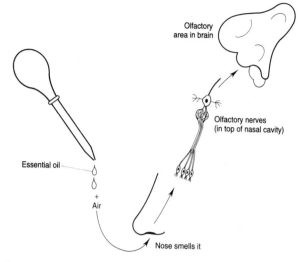

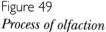

Figure 49
Process of olfaction

- nerve fibres of olfactory cells pass through a bony plate at the top of the nose and connect directly with the area of the brain known as the olfactory bulb

- smell is perceived by olfactory cells which connect directly with olfactory bulb in the brain

---

━━━━━ KEY NOTE ━━━━━

In most nerves in the body, the transmission of a nerve impulse is achieved through the spinal cord and then on to the brain. However, in the case of the olfactory cells, the nerve fibres connect *directly* with the olfactory bulb of the brain and therefore have a powerful and immediate effect on the emotions.

This can be explained by the fact that the area of the brain associated with smell is very closely connected with that part of the brain known as the limbic system, which is concerned with emotions, memory and sex drive. The olfactory bulb also connects closely with the hypothalamus, the nerve centre which governs the endocrine system.

Essential oils enter the nose in the form of gases, as they evaporate when in contact with air and are volatile in nature. It is in this evaporated form that we inhale them.

? SELF-ASSESSMENT QUESTIONS

1. List the individual parts which constitute the olfactory system

....

....

....

....

....

....

2. What is the functional significance of the olfactory cells and where are they located?

....

....

....

....

....

3. Briefly describe the process of olfaction

....

....

....

....

....

....

....

....

Chapter 11

THE NERVOUS SYSTEM

The nervous system is the main communication system for the body and is therefore responsible for receiving and interpreting information from inside and outside the body. The nervous system works intimately with the endocrine system to help regulate body processes.

A competent therapist needs to be able to understand the principles of operation of the nervous system to be able to carry out treatments safely and effectively.

By the end of this chapter, you will be able to relate the following information to your practical work carried out in the salon:

- The types of nerve impulses
- The transmission of nerve impulses
- The function of a motor nerve in muscle contraction
- The organisation of the nervous system
- An outline of the principal parts of the nervous system

THE ORGANISATION OF THE NERVOUS SYSTEM

The nervous system has two main divisions:

- The **central nervous system** consisting of the **brain** and the **spinal cord**
- The **peripheral nervous system,** which consists of thirty-one pairs of **spinal nerves**, twelve pairs of **cranial nerves** and the **autonomic nervous system.**

The functional unit of the nervous system is a *neurone* which is a specialised nerve cell, designed to receive stimuli and conduct impulses. Each neurone has a cell body, a central nucleus and processes called axons and dendrites which are extensions of the nerve cell. The function of a neurone is to transmit nerve impulses.

There are two main types of nerve impulses:

- **sensory** or *afferent* neurones receive stimuli from sensory organs and receptors and transmit the impulse to the spinal cord and brain. Sensations transmitted by the sensory neurones include heat, cold, pain, taste, smell, sight and hearing.

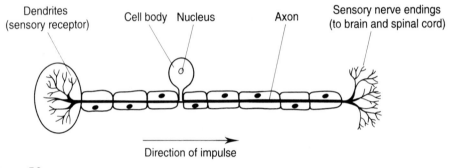

Figure 50
Sensory nerve

- **motor** or *efferent* neurones conduct impulses away from the brain and the spinal cord to muscles and glands in order to stimulate them into carrying out their activities.

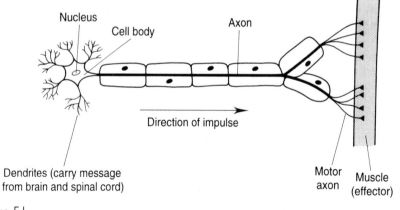

Figure 51
Motor nerve

THE TRANSMISSION OF NERVE IMPULSES

The function of a neurone is to transmit impulses from their origin to destination. The nerve fibres of a neurone are not actually joined together and therefore there is no anatomical continuity between one neurone and another. The junction where nerve impulses are transmitted from one neurone to another is called a synapse, which is a minute gap between the nerve fibres.

Impulses are relayed from one neurone to another by a chemical transmitter substance, which is released by the neurone to carry impulses across the synapse to stimulate the next neurone. Synapses cause nerve impulses to pass in one direction only and are important in co-ordinating the actions of neurones. A special kind of synapse occurs at the junction between a nerve and a muscle and is known as a motor point, which is the point where the nerve supply enters the muscle.

The conduction of motor impulse in the contraction of voluntary muscle:

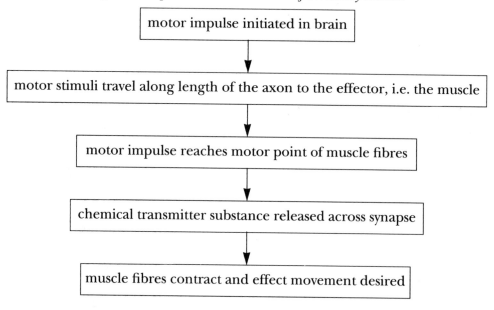

THE ORGANISATION OF THE NERVOUS SYSTEM

The central nervous system consists of two parts:

* the brain
* the spinal cord

THE BRAIN

The human brain is an extremely complex mass of nervous tissue which lies within the skull. It is the main communication centre of the nervous system and its function is to co-ordinate stimuli received and effect the correct responses.

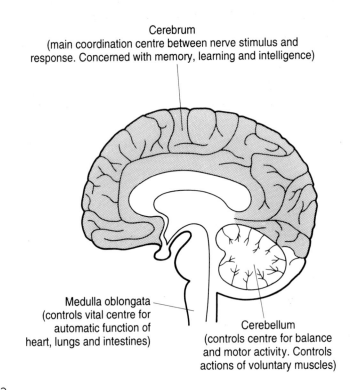

Cerebrum
(main coordination centre between nerve stimulus and response. Concerned with memory, learning and intelligence)

Medulla oblongata
(controls vital centre for automatic function of heart, lungs and intestines)

Cerebellum
(controls centre for balance and motor activity. Controls actions of voluntary muscles)

Figure 52
The brain

THE SPINAL CORD

The spinal cord extends from the large opening at the base of the skull down to the second lumbar vertebra and is an extension of the brain stem or medulla oblongata.

THE FUNCTIONS OF THE SPINAL CORD

- to relay impulses to and from the brain; sensory tracts conduct impulses to the brain and motor tracts conduct impulses from the brain
- provides the nervous link between the brain and other organs of the body

• is the centre for reflex actions which provide a fast response to external or internal stimuli

---

KEY NOTE

A reflex action is a rapid and automatic response to a stimulus, without any conscious thought of the brain. A typical example of a reflex action is the knee jerk response, which involves sensory and motor nerves being co-ordinated through the spinal cord.

THE PERIPHERAL NERVOUS SYSTEM

This part of the nervous system consists of:

• 31 pairs of spinal nerves
• 12 pairs of cranial nerves
• the autonomic nervous system

The peripheral nervous system is composed of the parts of the nervous system outside the brain and the spinal cord.

THE CRANIAL NERVES

The 12 pairs of cranial nerves connect directly to the brain and between them they provide a nerve supply to sensory organs, muscles and skin of the head and the neck. The facial nerves of interest to the therapist are as follows:

• trigemenal – fifth cranial nerve
• facial – seventh cranial nerve
• accessory – accessory cranial nerve

THE SPINAL NERVES

There are 31 pairs of spinal nerves which pass out of the spinal cord and each has two thin branches which link it with the autonomic nervous system. Spinal nerves receive sensory impulses from the body and transmit motor signals to specific regions of the body. Each of the spinal nerves are numbered and are named according to the level of the spinal column from which they emerge. They are as follows:

• eight cervical
• twelve thoracic
• five lumbar

- five sacral
- one coccygeal

Each spinal nerve is divided into several branches, forming a network of nerves, or plexuses which supply different parts of the body:

- The cervical plexuses of the neck supply the skin and muscles of the head, neck and upper region of the shoulders
- The brachial plexuses supply the skin and the muscles of the arm, shoulder and upper chest
- The lumbar plexuses supply the front and the sides of the abdominal wall and part of the thigh
- The sacral plexuses at the base of the abdomen supply the skin and the muscles and organs of the pelvis
- The coccygeal plexus: this is a small plexus supplying the skin in the area of the coccyx and the muscles of the pelvic floor

THE AUTONOMIC NERVOUS SYSTEM

This is the involuntary part of the nervous system which controls the automatic body activities of smooth and cardiac muscle and the activities of glands. The autonomic system is divided into two parts:

The **sympathetic** and **parasympathetic** nervous systems

This **sympathetic stimulation** produces a series of changes which prepare the body for vigorous activity, i.e. the so-called fight or flight syndrome. The effects are as follows:

- increased cardiac output, therefore the heart beats faster
- the bronchi dilate, increasing ventilation rate
- the skeletal blood vessels dilate
- the adrenals and sweat glands secrete abundantly
- the salivary and digestive glands secrete less, causing digestion to slow down
- liver produces more glucose and releases it
- the iris of the pupils dilate

--------- **KEY NOTE** ---------

The sympathetic stimulation of the autonomic system is increased by the release of the hormone adrenaline from the adrenal medulla, an example of the nervous and endocrine systems working in harmony with one another.

The **parasympathetic stimulation** has an inhibitory effect, preparing the body for inactivity, and works towards the conservation of the body and the restoration of energy. It therefore creates the conditions needed for rest, sleep and slows down the body processes except digestion and the functions of the genito-urinary system. In general its actions oppose those of the sympathetic system and the two systems work together to regulate the internal workings of the body. The effects of parasympathetic stimulation are as follows:

- heart beat slows down
- bronchi constrict
- skeletal blood vessels constrict
- peristalsis is increased and the secretion of insulin and digestive juices are increased
- pupils constrict

The sympathetic and parasympathetic systems are finely balanced to ensure the optimum functioning of organs.

TASK I

Complete the following table showing the differences in effects of the sympathetic and parasympathetic systems.

TABLE 22: *Sympathetic and parasympathetic nervous system*

Organ	Sympathetic stimulation	Parasympathetic stimulation
heart		
skeletal blood vessels		
bronchi		
pupils of eyes		
digestive		

? SELF-ASSESSMENT QUESTIONS

1. What is the function of a nerve?

....

....

....

2. What is the difference between a sensory and a motor nerve?

....

....

....

....

....

....

3. What is a motor point?

....

....

....

....

4. Briefly describe how a motor impulse is involved in the contraction of voluntary muscle

....

....

....

....

....

5. How is the nervous system organised? List the parts

....

....

....

Chapter 12

〜

THE ENDOCRINE SYSTEM

The endocrine system comprises a series of internal secretions called *hormones* which help to regulate body processes by providing a constant internal environment. The endocrine system works closely with the nervous system; nerves enable the body to respond rapidly to stimuli, whereas the endocrine system causes slower and longer lasting effects.

A competent therapist needs to be able understand the action of hormones and their significance in the healthy functioning of the body, as over secretion and under secretion of hormones may result in certain disorders and disease.

By the end of this chapter, you will be able to relate the following knowledge to your practical work carried out in the salon:

- what a hormone is
- the location of the main endocrine glands of the body
- the principal secretions of the main endocrine glands
- the effects of hormones on the body

WHAT IS A HORMONE?

A hormone is a chemical messenger or regulator, secreted by an endocrine gland, which reaches its destination by the bloodstream, and has the power of influencing the activity of other organs. Some hormones have a slow action over a period of years, for instance the growth hormone from the anterior pituitary, while others have a quick action such as adrenaline from the adrenal medulla. Hormones therefore regulate and co-ordinate various functions in the body.

The endocrine glands are ductless glands as the hormones they secrete pass directly into the bloodstream to influence the activity of another organ or gland. The main endocrine glands are as follows:

- the pituitary gland
- the thyroid gland
- the parathyroid glands
- the adrenal glands
- the islets of langerhans
- ovaries in the female
- testes in the male

THE PITUITARY GLAND

This is a lobed structure attached by a stalk to the hypothalamus of the brain. It is often referred to as the 'master gland' since it produces several hormones or *releasing factors* which influence the secretion of hormones by other endocrine organs. The pituitary gland consists of two main parts, an anterior and a posterior lobe.

THE ANTERIOR LOBE OF THE PITUITARY GLAND

The principal hormones secreted by the anterior lobe of the pituitary are as follows:

- **growth hormone**, which controls the growth of long bones and muscles.
- **thyroid stimulating hormone (TSH)**, which controls the growth and activity of the thyroid gland
- **adrenocorticothrophic hormone (ACTH)**, which stimulates and controls the growth and hormonal output of the adrenal cortex
- **gonadotrophic hormones** control the development and growth of the ovaries and testes. The gonads or sex hormones include:
  follicle stimulating hormone, which in women stimulates the development of the graafian follicle in the ovary which secretes the hormone oestrogen. In men it stimulates the testes to produce sperm.
  luteinizing hormone, which in women helps to prepare the uterus for the fertilised ovum. In men, it acts on the testes to produce testosterone.
- **prolactin** stimulates the secretion of milk from the breasts following birth

---
KEY NOTE

Endocrine glands in the body have a feedback mechanism which is co-ordinated by the pituitary gland. The pituitary gland is influenced by the hypothalamus and will increase its output of releasing factors if other glands start to fail or will decrease its output if the level of the hormone in the bloodstream starts to rise.

THE POSTERIOR LOBE OF THE PITUITARY GLAND

The posterior lobe of the pituitary secretes two hormones, which are manufactured in the hypothalamus but are stored in the posterior lobe:

- **the anti-diuretic hormone (ADH)** which increases water re-absorption in the renal tubules of the kidneys
- **oxytocin** stimulates the uterus during labour and stimulates the breasts to produce milk

THE THYROID GLAND

The thyroid gland is found in the neck, situated on either side of the trachea and is controlled by the anterior lobe of the pituitary. The principal secretion from the thyroid gland is the hormone **thyroxin**, which is intimately concerned with controlling metabolism. The functions of the thyroid gland are as follows:

- controls the metabolic rate by stimulating metabolism
- influences growth and cell division
- influences mental development
- is responsible for the maintenance of healthy skin and hair
- stores the mineral iodine, which it needs to manufacture thyroxin
- stimulates the involuntary nervous system and controls irritability

The thyroid gland is controlled by a feedback mechanism and will increase to meet the demand for more thyroid hormones at various times, such as during the menstrual cycle, at pregnancy, puberty.

THE PARATHYROID GLANDS

These are four small glands situated on the posterior of the thyroid gland. Their principal secretion is the hormone **parathormone** which helps to regulate calcium metabolism by controlling the amount of calcium in blood and bones.

THE ADRENAL GLANDS

These are two triangular shaped glands which lie on top of each kidney. They consist of two parts, an outer cortex and an inner medulla.

THE ADRENAL CORTEX

The principal hormones secreted by the adrenal cortex are as follows:

- **gluco corticoids**, which are hormones that influence the metabolism of protein and carbohydrates and the utilisation of fats
- **mineral corticoids**, which act on the kidneys, helping to maintain the water and mineral balance in the body
- **sex corticoids**, which control the development of the secondary sex characteristics and the function of the reproductive organs

— K E Y N O T E —

When the ovaries and testes mature, they produce the sex hormones themselves, therefore the production of sex corticoids in the adrenal cortex is important up to puberty.

THE ADRENAL MEDULLA

The principal hormone secreted by the adrenal medulla is **adrenaline**, which is under the control of the sympathetic nervous system and is released at times of stress. The response of this hormone is fast due to the fact that it is governed by nervous control. The effects of adrenaline on the body are as follows:

- dilates the arteries, increasing blood circulation and the rate of heart beat
- dilates the bronchial tubes, increasing oxygen intake and the rate and depth of breathing
- constricts blood vessels to the skin and the intestines, diverting blood from these regions to your muscles and brain to effect action
- reduces digestion
- increases activity of the sweat glands

— K E Y N O T E —

The effects described above are those felt when the body is under stress – a pounding heart, increased ventilation rate, a dry mouth and butterflies in the stomach.

THE PANCREAS

The pancreas is known as a dual organ, as it has an endocrine and an exocrine function:

- The exocrine or external secretion is the secretion of pancreatic juice, to assist with digestion
- The endocrine or internal secretion is the hormone insulin, secreted by the islets of langerhans cells in the pancreas. Insulin lowers the level of sugar in the blood by helping the body cells to take it up and use or store it as glycogen

THE SEX GLANDS

THE TESTES

The testes are situated in the groin, in a sac called the scrotum. They have two functions:

- the secretion of the hormone **testosterone**, which controls the development of the secondary sex characteristics in the male at puberty (influenced by the luteinising hormone)
- the production of **sperm** (influenced by the follicle stimulating hormone from the anterior pituitary)

THE OVARIES

The ovaries are situated in the lower abdomen, below the kidneys. The two ovaries are the sex glands in the female, each is attached to the upper part of the uterus by broad ligaments. The ovaries have two distinct functions:

- the production of ova at ovulation
- production of the two hormones oestrogen and progesterone, which influence the secondary sex characteristics in the female and affect the process of reproduction

PUBERTY

The hormones oestrogen and progesterone become active at puberty and are responsible for the development of the secondary sex characteristics. The ovaries are stimulated by the gonadotrophic hormones from the anterior pituitary, known as the follicle stimulating hormone and the luteinising hormone. At puberty, oestrogen causes the breasts, the vulva and the vagina to gradually develop to mature proportions.

Oestrogen influences the amount of subcutaneous fat, determining the rounded contours of the female figure and is also concerned with the growth of the milk ducts in the breast. Oestrogen also influences the menstrual cycle and thickens the uterus lining in preparation for conception.

KEY NOTE

Menstrual disorders occur when the ovaries fail to respond to the stimulation by the gonadotrophic hormones from the anterior pituitary or when there is an abnormal response to stimulation. This then results in increased production of male hormones called androgens, which may lead to hair growth in the male sexual pattern, a condition known as hirsutism.

PREGNANCY

Progesterone is known as the pregnancy hormone as it is concerned with the development of the placenta, which is a temporary endocrine gland during pregnancy. Progesterone helps with maintenance of the pregnancy and prepares the breasts for lactation.

MENOPAUSE

The menopause marks the end of a woman's reproductive life when oestrogen production begins to decline. The ovaries gradually become less responsive to the sex hormones and hence ovulation and the menstrual cycles become irregular until they cease altogether.

KEY NOTE

Normally, hormones produced by the ovaries have an inhibitory or restraining effect on the anterior pituitary. However, in the case of the menopause, a lack of oestrogen results in a lack of proper control over this master gland, which then begins to pour out a flood of stimulating hormones.

This results in hyper stimulation by the pituitary hormones of the adrenal cortex, which in turn produces an excess of androgens or male hormones. It is for this reason that women of menopausal age, whose ovarian activity is declining, find themselves developing excess facial and body hair.

MAIN ENDOCRINE GLANDS IN THE BODY

☞ TASK 1

Label the main endocrine glands of the body on the following diagram.

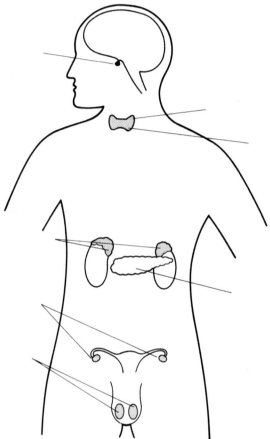

Figure 53
Main endocrine glands

THE HORMONAL SECRETIONS OF THE MAIN ENDOCRINE GLANDS

☞ TASK 2

Complete Table 23 by identifying the name of the hormone and the endocrine gland it is secreted from.

TABLE 23: *Main endocrine glands and their hormonal secretions*

Name of hormone	Secreted by	Effects on body
		controls growth and development of long bones and muscle
		controls growth and development of ovaries and testes
		stimulates secretion of milk from breasts after birth
		stimulates uterus in labour, stimulates breasts to produce milk
		controls metabolism
		regulates calcium metabolism
		helps maintain water and mineral balance in kidneys
		prepares body for stressful situation, e.g. fight or flight
		controls metabolism of glucose
		production of ova, responsible for development of secondary sex characteristics in female
		prepares uterus for pregnancy
		controls secondary sex characteristics in male, stimulates production of sperm

? SELF-ASSESSMENT QUESTIONS

1. What is a hormone?

....

....

....

....

....

....

2. Name the main endocrine glands in the body

....

....

....

....

....

....

....

....

3. Why is the pituitary gland known as the master gland?

....

....

....

....

....

4. Describe the effects of the hormone adrenaline on the body

....

....

....

....

5. Briefly describe the effects the hormones oestrogen and progesterone have on the body at:
   i) puberty

....

....

....

....

ii) pregnancy

....

....

....

....

iii) menopause

....

....

....

....

6. What is meant by the term hirsutism and what may cause this to occur?

....

....

....

....

....

....

Chapter 13

THE FEMALE BREAST

The female breasts are accessory organs to the female reproductive system and their function is to produce and secrete milk after pregnancy. A competent therapist needs to have a basic knowledge of the structure of the breast to understand how salon treatments may affect the size and shape of the breast.

By the end of this chapter, you will be able to relate the following knowledge to your practical work carried out in the salon:

- brief structure of the breast
- the function of the breast
- factors affecting the size and shape of the breasts

POSITION

The breasts lie on the pectoral region of the front of the chest and are situated between the sternum and the axilla, extending from approximately the second to the sixth rib. The breasts lie over the pectoralis major and serratus anterior muscles and are attached to them by a layer of connective tissue.

STRUCTURE

The breasts consist of glandular tissue arranged in lobules, supported by connective, fibrous and adipose tissue. The lobes are divided into lobules which open up into milk ducts.

The milk ducts open into the surface of the breast at a projection called the nipple. Around each nipple, the skin is pigmented and forms the areola, which varies in colour from a deep pink to a light or dark brown colour. A considerable amount of fat or adipose tissue covers the surface of the gland and is found between the lobes. The skin on the breast is thinner and more translucent than the body skin.

SUPPORT

The breasts are supported and slung in powerful suspensory cooper's ligaments, which go around the breast, both ends being attached to the chest wall. The pectoralis major and serratus anterior muscles help to support the ligaments.

If the breast grows large due to adolescence or pregnancy, the cooper's ligaments may become irreparably stretched and the breast will then sag. With age, the supporting ligaments, along with the skin and the breast tissue, become thin and inelastic and the breasts lose their support.

LYMPHATIC DRAINAGE OF THE BREAST

The breasts contain many lymphatic vessels and the lymph drainage is very extensive, draining mainly into the axillary nodes under the arms.

BLOOD SUPPLY OF THE BREAST

The blood vessels supplying blood to the breast include the subclavian and axillary arteries.

NERVE SUPPLY

There are numerous sensory nerve endings in the breast, especially around the nipple. When these touch receptors are stimulated in lactation, the impulses pass to the hypothalamus and the flow of the hormone oxytocin is increased from the posterior lobe of the pituitary, promoting the constant flow of milk when required.

HORMONES RESPONSIBLE FOR THE DEVELOPMENT OF THE BREAST

- **Oestrogen**: is responsible for the growth and development of the secondary sex characteristics

- **Progesterone**: causes the mammary glands to increase in size if fertilisation and subsequent pregnancy occurs.

DEVELOPMENT OF THE BREASTS

PUBERTY

The breast start out as a nipple which projects from the surrounding ring of pigmented skin called the areola. Approximately two or three years before the onset of menstruation, the fat cells enlarge in response to the sex hormones, oestrogen and progesterone, released during adolescence.

———————— **K E Y N O T E** ————————

The breasts change monthly in response to the menstrual cycle. The action of the female hormone progesterone increases blood flow to the breast which increases fluid retention, and the breast may increase in size causing them to feel swollen and uncomfortable.

PREGNANCY

During pregnancy, the increased production of oestrogen and progesterone causes an increase in blood flow to the breast. This causes an enlargement of the ducts and lobules of the breast in preparation for lactation, and there is an increase in fluid retention. The areola and the nipple enlarge and become more pigmented.

MENOPAUSE

The reduction in the female hormones during the menopause causes the glandular tissue of the breast to shrink and the supporting ligaments, along with the skin, become thinner and lose their elasticity. Therefore, during the menopause the breasts begin to lose their support and uplift, although the degree of loss is dependent on the original strength of the suspensory ligaments.

FACTORS DETERMINING SIZE AND SHAPE

The size of the breast is largely determined by genetic factors, although there are other factors such as:

- amount of adipose tissue present
- fluid retention

- level of ovarian hormones in the blood and the sensitivity of the breasts to these hormones
- degree of ligamentary suspension
- exercise undertaken

KEY NOTE

Exercise may help to strengthen the pectoral muscles which will help to support the ligaments and increase their uplift. However, if the wrong type of exercise is undertaken and insufficient support is not provided for the breasts during exercise, the ligaments may become irreparably stretched.

THE STRUCTURE OF THE FEMALE BREAST

◆◆ TASK I

Label the main structures that form the female breast.

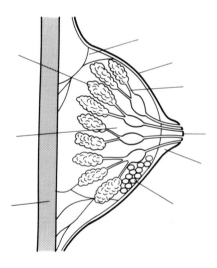

Figure 54
Cross section of the breast

? SELF-ASSESSMENT QUESTIONS

1. What is the anatomical position of the breast?

....

2. Name the main type of tissue which makes up the breast

....

....

3. Name the hormones responsible for the development of the breasts

....

....

....

4. List the changes which occur to the breast during
   i) puberty

....

....

....

....

ii) pregnancy

....

....

....

....

iii) menopause

....

....

....

....

5. List four factors affecting the shape and size of the breast

....

....

....

....

Chapter 14

THE DIGESTIVE SYSTEM

The digestive system involves food being broken down and made soluble before it can be absorbed by the body for nutrition. Food is taken in through the mouth, broken into smaller particles and absorbed into the bloodstream, where it is utilised by the body. Waste materials not required by the body are then passed through the body to be eliminated. Once food has been absorbed by the body, it is converted into energy to fuel the body's activities; this is known as metabolism.

A competent therapist needs to be able to have knowledge of the process of the breakdown of food in order to understand how the body utilises nutrients for efficient body functioning.

By the end of this chapter, you will be able to relate the following knowledge to your work carried out in the salon:

- the process of digestion from the ingestion of food to the elimination of waste
- the structure and functions of the organs associated with digestion
- the absorption of nutrients and their utilisation in the body
- sources and functions of the main food groups required for good health

Digestion occurs in the alimentary tract, which is a long continuous muscular tube, extending from the mouth to the anus. It consists of the following parts:

- the mouth
- the pharynx
- the oesophagus
- the stomach

- the small intestine
- the large intestine
- the anus

The *pancreas, gall bladder* and the *liver* are accessory organs to digestion.

MOUTH

The digestive system commences in the mouth, where food is broken up into smaller pieces by the teeth, is shaped into a ball by the tongue and is mixed thoroughly with saliva from the salivary glands, which open into the mouth. The digestion of *starch* commences in the mouth.

PHARYNX AND OESOPHAGUS

The ball of food is projected to the back of the mouth and the muscles of the pharynx force the food down the oesophagus, which is a long narrow tube linking the pharynx to the stomach. The food is then conveyed by peristalsis down the oesophagus to the stomach.

— K E Y   N O T E —

Peristalsis is the co-ordinated rhythmical contractions of the circular and oblique muscles in the wall of the alimentary canal. These muscles work in opposition to one another to break food down and move it along the alimentary canal. Peristalsis is an automatic action stimulated by the presence of food and occurs in all sections of the alimentary canal.

DIGESTION IN THE STOMACH

The stomach is a curved muscular organ, positioned in the left hand side of the abdominal cavity, below the diaphragm. The stomach has:

- a serous membrane which prevents friction
- a muscular coat which assists the mechanical breakdown of food
- numerous gastric glands which secrete gastric juice
- a mucous coat which secretes mucous to protect the stomach lining from the damaging effects of the acidic gastric juice

The functions of the stomach are to:

- churn and to break up large particles of food mechanically
- mix food with gastric juice to begin the chemical breakdown of food
- commence the digestion of *protein*

Food stays in the stomach for approximately five hours, until it has been churned down to a liquid state called chyme. Chyme is then released at intervals into the first part of the small intestine.

DIGESTION IN THE SMALL INTESTINE

The small intestine is made up of the same coats as the stomach and consists of three parts:

- the duodenum, which is the first part of the small intestine
- the jejunum
- the ileum, where the main absorption of food takes place

Special features of the small intestine are the thousands of minute projections called *villi* into which the nutrients pass to be absorbed into the bloodstream. The muscles in the wall of the small intestine continue the mechanical breakdown of food by the action of peristalsis. The chemical breakdown of food is completed by the following juices, which prepare the food to be absorbed into the bloodstream.

- **Bile**, stored by the gall bladder, which is a muscular, membranous bag situated on the underside of the right lobe of the liver. Bile is an alkaline liquid consisting of water, mucus, bile pigments, bile salts and cholesterol, and is released at intervals from its duct when food enters the duodenum. The function of bile is to neutralise the chyme and break up any fat droplets in a process called emulsification
- **Pancreatic juice**, produced by the pancreas, which is a gland extending from the loop of the duodenum to behind the stomach. The pancreas secretes pancreatic juice into the duodenum and the enzymes contained within it continue the digestion of protein, carbohydrates and fat
- **Intestinal juice**, which is released by the glands of the small intestine and completes the final breakdown of nutrients, including simple sugars to glucose and protein to amino acids

ABSORPTION OF THE DIGESTED FOOD

The absorption of the digested food occurs by diffusion through the villi of the small intestine. The villi are well supplied with blood capillaries to allow the digested food to enter.

- Simple sugars from *carbohydrate* digestion and amino acids from *protein* digestion pass into the bloodstream via the villi and are then carried to the liver to be processed
- Products of *fat* digestion pass into the intestinal lymphatics which absorb the fat molecules and carry them through the lymphatic system before they reach the blood circulation
- *Vitamins* and *minerals* travel across to the blood capillaries of the villi and are absorbed into the bloodstream to assist in normal body functioning and cell metabolism.

THE LIVER

The liver is the largest gland in the body and is situated in the upper right hand side of the abdominal cavity, under the diaphragm. It is a vital organ and therefore has many important functions in the metabolism of food as it regulates the nutrients absorbed from the small intestine to make them suitable for use in the body's tissues. Its functions are:

- **Secretion of bile.** Bile is manufactured by the liver but is stored and released by the gall bladder to assist the body in the breakdown of fats
- **Regulation of blood sugar levels.** When the blood sugar levels rise after a meal, the liver cells store excess glucose as glycogen. Some glucose may be stored in the muscle cells as muscle glycogen
- When both these stores are full, surplus glucose is converted into fat by the liver cells
- **Regulation of amino acid levels.** As our bodies cannot store excess protein and amino acids, they are processed by the liver; some are removed by the liver cells and are used to make plasma proteins, some are left for the body cells tissues' use, whilst the rest are deaminated and excreted as urea in the kidneys
- **Regulation of the fat content of blood.** The liver is involved in the processing and transporting of fats; those already absorbed in the diet are used for energy, and excess fats are stored in the tissues.
- **Regulation of plasma proteins.** The liver is active in the breakdown of worn out red blood cells

- **Detoxification.** The liver detoxifies harmful toxic waste and drugs and excretes them in bile or through the kidneys
- **Storage.** The liver stores vitamins A, D, E, K and B12 and the minerals iron, potassium and copper
- The liver can also hold up to a litre of blood and during exercise the liver supplies extra blood and increases oxygen transport to the muscles

THE PRODUCTION OF HEAT

- due to its many functions the liver generates heat

Once all the nutrients have been absorbed into the bloodstream they are transported to the body's cells for metabolism:

- **Glucose**, which is the end product of carbohydrate digestion and is used to provide energy for the cells to function
- **Amino acids**, which are the end products of protein digestion and are used to produce new tissues, repair damaged cell parts and to formulate enzymes, plasma proteins and hormones
- **Fatty acids and glycerol** are the end products of fat digestion. Fats are used primarily to provide heat and energy, in addition to glucose. Those fats which are not required immediately by the body are used to build cell membranes, and some are stored under the skin or around vital organs such as the kidneys and the heart.

When all the body's nutrients have been assimilated by the body, the fate of the undigested food is to pass into the large intestine where it is eventually eliminated from the body.

THE LARGE INTESTINE

The large intestine coils around the small intestine and is made up of bands of longitudinal muscle and folds of mucosa which secrete mucous. The colon is the main part of the large intestine and has three bends of flexures:

- ascending
- transverse
- descending

The functions of the large intestine are:

- absorption of most of the water from the faeces in order to conserve moisture in the body

- formation and storage of faeces (which consists of undigested food, dead cells and bacteria)
- production of mucus to lubricate the passage of faeces
- the expulsion of faeces out of the body, through the anus

MAIN DIGESTIVE ORGANS

►► TASK I

Label the main digestive organs on the following diagram.

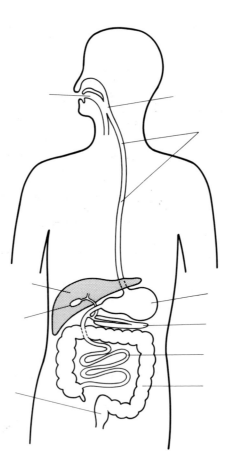

Figure 55
Outline of the digestive tract

● **TASK 2**

Complete the following table which summarises the functions of the main parts of the digestive tract.

TABLE 24: *Functions of main digestive organs*

Name of organ	Function of organ
	ingestion of food, mechanical breakdown of food by teeth
	swallowing
	links pharynx to stomach
	churns food, mechanical breakdown of food
	chemical breakdown of food, completion of digestion, absorption into bloodstream
	absorption of water formation and storage of faeces
	elimination of waste

NUTRITION – FOOD GROUPS

● **TASK 3**

An adequate diet must provide sufficient nutrients to support growth and development of the body.

Complete the following table to show the sources and functions of the main food groups necessary for good health.

TABLE 25: *Sources and functions of food groups*

Food group	Sources	Main functions
carbohydrates		
proteins		
fats		
water		
fibre		
vitamin A (fat soluble)		
vitamin D (fat soluble)		
vitamin E (fat soluble)		
vitamin K (fat soluble)		
vitamin B1 (water soluble)		
vitamin B2 (water soluble)		
vitamin B5 (water soluble)		

TABLE 25: *Continued*

Food group	Sources	Main functions
vitamin B6 (water soluble)		
vitamin B12 (water soluble)		
folic acid (water soluble)		
vitamin C (water soluble)		
calcium		
iron		
phosphorus		
sulphur		
sodium and chlorine		
magnesium		

? SELF-ASSESSMENT QUESTIONS

1. What is metabolism?

....

....

....

....

2. Give a brief account of the process of the breakdown of food, from ingestion to the elimination of waste.

....

....

....

....

....

....

....

....

....

....

3. List four metabolic functions of the liver

....

....

....

....

....

4. Briefly explain how the following nutrients are utilised by the body
   a) glucose

....

....

....

....

   b) amino acids

....

....

....

....

   c) fats

....

....

....

....

Chapter 15

THE RENAL SYSTEM

The renal system is made up of excretory organs which are involved in the processing and elimination of normal metabolic waste from the body. Waste products such are urea and uric acid, along with excess water and mineral salts must be removed from the body in order to maintain good health. If these waste materials were allowed to accumulate in the body they would cause ill health. The primary function of the renal system, therefore, is to regulate the composition and the volume of body fluids, in order to provide a constant internal environment for the body.

A competent therapist needs to be able to have knowledge of the outline of the renal system to understand how fluid balance is controlled in the body.

By the end of this chapter, you will be able to relate the following knowledge to your practical work carried out in the salon:

- the structure of the individual parts of the renal system
- the functions of the individual parts of the renal system
- the regulation of fluid balance in the body

The urinary system consists of the following parts:

- two **kidneys** which secrete urine
- two **ureters** which transport urine from the kidneys to the bladder
- one **urinary bladder** where urine collects and is temporarily stored
- one **urethra** through which urine is discharged from the bladder and out of the body

THE KIDNEYS

The kidneys lie on the posterior wall of the abdomen, on either side of the spine, between the level of the twelfth thoracic vertebrae and the third lumbar vertebrae. A kidney has an outer fibrous capsule and is supported by adipose tissue. It has two main parts:

- an outer *cortex*, which is the outermost section where the fluid is filtered from the blood
- an inner *medulla* where some materials are re-absorbed back into the bloodstream

The blood that needs to be processed enters the medulla from the renal artery. Inside the kidney the renal artery splits into a network of capillaries which filter the waste. Some substances contained within the waste such as glucose, amino acids, mineral salts and vitamins are re-absorbed back into the bloodstream, as the body cannot afford to lose them.

Excess water, salts and the waste product urea are all filtered and processed through the kidneys and the treated blood leaves the kidney via the renal vein. Urine, the waste product of filtration produced by the kidney, collects in a funnel-shaped structure called the renal pelvis, from which it flows into the ureter.

———— K E Y N O T E ————

Urine consists of water, salts and protein wastes and varies in its colour according to its composition and quantity. An analysis of the substances present in the urine is a good indication of the state of health of the body and urine tests are often used to diagnose disorders and diseases in the body.

FUNCTION

The functions of the kidney are:

- filtration of impurities and metabolic waste from blood, and preventing poisons from fatally accumulating in the body
- regulation of water and salt balance in the body
- maintenance of the normal pH balance of blood
- formation of urine

THE ROLE OF THE KIDNEYS IN FLUID BALANCE

The amount of fluid taken into the body must equal the amount of fluid excreted from it in order for the body to maintain a constant internal environment. The balance between *water intake* and *water output* is controlled by the kidneys.

Water intake

Water is mainly taken into the body as liquid through the process of digestion; however, some is also released through the cell's metabolic activities.

Water output

Water is lost from the body in the following ways:

- through the kidneys as urine
- through the alimentary tract as faeces
- through the skin as sweat
- through the lungs as saturated exhaled breath

The kidneys are responsible for regulating the amount of water contained within the blood:

- If you have an excess of water in the blood, the blood concentration will be dilute, and the nerve receptors in the hypothalamus will trigger the pituitary gland to send a message to the kidneys to reduce water reabsorption, in order that a more dilute urine is eliminated from the body.

This mechanism reduces the amount of water in the blood back to an acceptable level

- If your blood is too concentrated and you do not have enough water, the nerve receptors in the hypothalamus trigger the pituitary gland to send a message to the kidneys to increase water re-absorption and a more concentrated urine is produced.

This increases the water content of blood back to an acceptable level and helps to conserve water in the body.

─────── KEY NOTE ───────

This important feedback mechanism between the nervous and endocrine systems maintains the blood concentration within normal limits and is the means by which fluid balance is controlled in the body.

Factors affecting fluid balance in the body include:

- **body temperature**: if the body temperature increases, more water is lost from the body in sweat
- **diet**: a high salt intake can result in increased water re-absorption which reduces the volume of urine produced. Diuretics such as alcohol, tea and coffee can also increase the volume of urine
- **emotions**: nervousness can result in an increased production of urine
- **blood pressure**: when the blood pressure inside the kidney tubules rises, less water is re-absorbed and the volume of urine will be decreased. When the blood pressure inside the kidney tubules falls, more water is re-absorbed into the blood and the volume of urine will be reduced.

THE URETERS

The ureters are two very fine muscular tubes which transport urine from the pelvis of the kidney to the urinary bladder. They consist of three layers of tissue:

- an outer layer of fibrous tissue
- a middle layer of smooth muscles
- an inner layer of mucous membrane

FUNCTION

- to propel urine from the kidneys into the bladder by the peristaltic contraction of their muscular walls.

THE URINARY BLADDER

This is a pear-shaped sac which lies in the pelvic cavity, behind the symphysis pubis. The size of the bladder varies according to the amount of urine it contains. The bladder is composed of four layers of tissue:

- a serous membrane which covers the upper surface
- a layer of smooth muscular fibres
- a layer of adipose tissue
- an inner lining of mucous membrane

FUNCTIONS

- storage of urine
- expels urine out of the body, assisted by the muscular wall of the bladder and the lowering of the diaphragm and the contraction of the abdominal cavity

THE URETHRA

This is a canal which extends from the neck of the bladder to the outside of the body. The length of the urethra differs in males and females; the female urethra being approximately only four centimetres in length, whereas the male urethra is longer, at approximately 18 to 20 centimetres in length. The exit from the bladder is guarded by a round sphincter of muscles which must relax before urine can be expelled from the body.

The urethra is composed of three layers of tissue:

- a muscular coat, continuous with that of the bladder
- a thin spongy coat, which contains a large number of blood vessels
- a lining of mucous membrane

FUNCTION

- serves as a tube through which urine is discharged from the bladder to the exterior
- as the urethra is longer in a male it also serves as a conducting channel for semen

MAIN RENAL ORGANS

☞ TASK 1

Label the main renal organs on the following diagram.

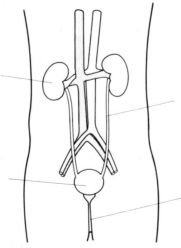

Figure 56
Organs of the renal system

❓ SELF-ASSESSMENT QUESTIONS

1. Briefly describe the basic structure of the kidneys

   ....

   ....

   ....

   ....

   ....

   ....

2. List the functions of the kidneys

   ....

   ....

   ....

   ....

   ....

   ....

   ....

   ....

3. What is the function of the ureter?

   ....

   ....

   ....

   ....

   ....

4. What is the function of the urinary bladder?

....

....

....

....

....

....

5. Briefly describe how fluid balance is controlled in the body

....

....

....

....

....

....

....

6. List two factors affecting fluid balance in the body

....

....

....

....

7. Name two excretory organs other than the kidneys

....

....

....

Candidate competency record
Anatomy and Physiology – Beauty Therapy Basics
for NVQ Levels II and III

Chapter 1 The skin	Date competency achieved	Assessor's signature
Task 1		
Task 2		
Task 3		
Self-assessment questions		
Chapter 2 The hair	Date competency achieved	Assessor's signature
Task 1		
Task 2		
Self-assessment questions		
Chapter 3 The nail	Date competency achieved	Assessor's signature
Task 1		
Task 2		
Task 3		
Task 4		
Self-assessment questions		
Chapter 4 The skeletal system	Date competency achieved	Assessor's signature
Task 1		
Task 2		
Task 3		
Task 4		
Self-assessment questions		
Chapter 5 The joints	Date competency achieved	Assessor's signature
Task 1		
Task 2		
Self-assessment questions		

Chapter 6 **The muscular system**	Date competency achieved	Assessor's signature
Self-assessment questions A		
Task 1		
Task 2		
Self-assessment questions B		
Task 3		
Task 4		
Task 5		
Task 6		
Task 7		
Task 8		
Task 9		
Task 10		
Task 11		
Task 12		
Task 13		
Task 14		
Task 15		
Task 16		
Task 17		
Self-assessment questions C		
Chapter 7 **The circulatory system**	Date competency achieved	Assessor's signature
Task 1		
Task 2		
Task 3		
Self-assessment questions		

Chapter 8 **The lymphatic system**	Date competency achieved	Assessor's signature
Task 1		
Task 2		
Task 3		
Self-assessment questions		
Chapter 9 **The respiratory system**	Date competency achieved	Assessor's signature
Task 1		
Self-assessment questions		
Chapter 10 **The olfactory system**	Date competency achieved	Assessor's signature
Self-assessment questions		
Chapter 11 **The nervous system**	Date competency achieved	Assessor's signature
Task 1		
Self-assessment questions		
Chapter 12 **The endocrine system**	Date competency achieved	Assessor's signature
Task 1		
Task 2		
Self-assessment questions		
Chapter 13 **The female breast**	Date competency achieved	Assessor's signature
Task 1		
Self-assessment questions		
Chapter 14 **The digestive system**	Date competency achieved	Assessor's signature
Task 1		
Task 2		
Task 3		
Self-assessment questions		

Chapter 15 The renal system	Date competency achieved	Assessor's signature
Task 1		
Self-assessment questions		

BIBLIOGRAPHY

Beckett, B. S. (1990) *Illustrated Human and Social Biology*, Oxford University Press.

Bennett, R. (1995) *The Science of Beauty Therapy*, London: Hodder & Stoughton.

Gaudin, A. J. and Jones, K. C. (1989) *Human Anatomy and Physiology*, San Diego, CA: Harcourt Brace Jovanovich.

Hole, J. W. Jr (1993) *Human Anatomy and Physiology*, Wm. C. Brown.

Moffatt, D. B. and Mottram, R. F. (1983) *Anatomy and Physiology for Physiotherapists*, Oxford; Blackwell Scientific Publications.

Pearce, E. (1983) *Anatomy and Physiology for Nurses*, London: Faber & Faber.

Rowett, H. G. (1983) *Basic Anatomy and Physiology*, London: John Murray.

Tortora and Grabowski (1992) *Principles of Anatomy and Physiology*, London: Harper Collins.

Wilson, K. J. W. (1990) *Anatomy and Physiology in Health and Illness*, Edinburgh: Churchill Livingstone.

Wright, D. (1983) *Human Biology*, Oxford: Heinemann Educational.